Fall Fever

Told by "Mister Trump" with humor and wit.
Just the way you remember him!

Richard F. Trump Sr.

AniLibre.com

Title: Fall Fever

Author: Richard F. Trump Sr.

Publisher: AniLibre.com
A Division of Aniable LLC
Fort Collins, Colorado

Second Edition, February 2, 2015

Persons named in this book are real people. Teachers and administrators are identified by their full names. Student are named with their first name and last initial. If the narrative demands a full name, it has been altered. If you recognize yourself or anyone else in these chapters, consider it an honor that "Mister Trump" remembered you! Feel free to take a pen and substitute your real name into the copy you'll give your grandkids!

Dedication by Richard F. Trump

To Lorene, with whom I shared this journey

Dedication by David M. Trump

To our Dad: March 19, 1912 – November 1, 2010 - Teacher

ISBN: **0692362207**
ISBN-13: **978-0692362204**

TABLE OF CONTENTS

Forward

By the time most people have made it all the way through their formal education they can name some "favorite teachers." I had several who inspired and enlightened me but my favorite teacher of all was "Mister Trump." Richard F. Trump, he would be on legal documents. Dick, to his teaching cohorts, Rich, to his wife of 62 years, but I just called him Dad.

In a preface to another of his books, I referred to him as "my Father." He objected. It was too formal, he said. But I had used that term out of immense respect. I wish for everyone the privilege of having a father, a dad, who deserves the degree of respect in which I hold my own. My Dad was a teacher, and I can think of no higher complement than that. It has been an honor to be his student for all of my life. In the "Monday" of this book, with his characteristic wry humor, you'll learn why, from the beginning of his teaching career, Dad insisted that his students call him "Mister Trump." By the time he left formal teaching 51 years later, he admits that the same reason no longer applied.

When I first wrote this forward for an earlier edition of this book, Dad was 96, and still a teacher. His newest students were folks at the senior living center where he then made his home. Whenever he had the opportunity, I'm sure that he still pointed out why the juncos come in large groups to the tree just outside the lounge area, or how the small pond just down the hill is gradually changing into a marsh. He was still my teacher, too.

I'm not sure if I was a good enough student of his in 1955 to have learned what he was trying to teach the rest of the sophomore biology class. But I had a head start - and I had another 55 years to continue to learn the most important thing that my Dad wanted to teach his students. Not the "facts" of life science, but the learning of life as a science. On "Wednesday", in "The Danger of Being Right" he talks about how he learned to value what became the most important principle of his career as a scientist. Later on the same "day" he mentions the idea again in "Training the Truth Searchers."

On "Thursday" he expresses the regret that many of his sophomores believe that the type of thinking required for his biology class is useful only in the study of science. Dad taught science, not as a way for his students to accumulate facts, but as a way of thinking. Read this book with that in mind and you will gain or strengthen an important skill.

But this is not a dry treatise on the scientific method, nor on the teaching of science. This is a narrative of a master teacher devoted to his subject, and devoted to his students - all of us - even more. He could laugh at us and he could laugh at himself.

Dad is gone now, but his writing is still with us. It is my privilege to make it available. It's a remembrance of a long career trying to instill a love of learning in young high school students. But I believe that it can be an inspiration and a delight to those who carry on the noblest profession: teacher.

David Trump, September 2, 2008 Revised February 2, 2015

Introduction

Late August, 1977. If I were a chipmunk, this feeling would send me off to the burr oak to harvest acorns. Being related to the chipmunk, I glance at the stack of fireplace fuel on the terrace and decide that one day soon I must sharpen my ax.

But my real concern is not shared by the chipmunk. The yellowing of poplar leaves, the reddening of hawthorn fruits, the noisy conferences of grackles - these are a calendar that reminds me to search the Squaw Creek pools for whirligig beetles, and to verify that the blazing star is still blooming. It tells me to take stock of ailanthus trees where I can clip a few stems for the study of leaf scars.

Such tasks take hold of me every autumn, a prelude to the delights and disappointments that I find in being a biology teacher.

Each school year ends in a weariness that comes with the frustrations of almost any teaching assignment. But there is renewal in the summer sun and a change of duties. Then as the days become shorter, I feel this urge. It is a sort of "fall fever."

More than a hundred fresh sophomores; some will join me in the excitement of seeing blood flow through the embryonic heart of a chick; others will not. A few will share their satisfaction in finding that their own cells and those of the blazing star blooming on the prairie just outside our classroom windows have a common plan; others won't. A few, having learned to track down the name of a flower or bug with an identification key, will ask for more; many will decline.

Some I will come to know. Others will remain strangers.

Perhaps this fall fever conditions me for what is ahead, just as the chipmunk's anxiety over the autumn harvest conditions it for winter. Whatever it is, I am glad it is a part of my life. It has been that way for 41 years; and even though retirement is providing time for sorting through the log of what happened, I find myself writing as though it is still happening.

It started in September, 1936, the MONDAY of my career, at Keokuk Senior High in the southeast corner of Iowa. TUESDAY begins in 1943 when we three - Lorene, three-year-old David, and I, moved to Ames, more nearly in the middle of the state. I will only mention briefly the two-year interlude when I was involved in World War II. The remaining three sections, WEDNESDAY through FRIDAY, bring me to 1977, in roughly chronological sequence.

While sorting and sifting the log, I will give my students fictitious names. This will be difficult; it will be prying away part of their identity. Perhaps, by that time, I will feel obliged to say what it is like to not be teaching. If so, it will become WEEKEND.

Monday

I soon recognized that it was a mistake to begin teaching in my home town.

Having started, I thought I should stay until I was sure of myself

It took seven years.

RETURN TO KHS

Keokuk Senior High School was in the middle of town. The two-tier yellow brick building faces Main Street, between 14th and 15th. Independent Lumber is across the street on 14th, Keasling Drug across Main. Blondeau Street runs parallel to Main, close to the back of the school. On the corner of 14th and Blondeau, old Joe Bailey still tends his little grocery store. Joe is a friend of boys. They buy candy, smoke cigarettes, and whistle to the girls who use the 14th Street door of the school.

Fifteen blocks down Main Street is the Mississippi River. My twelve-foot skiff is tied to the patchy dock that wobbles in the current below the dam. I built the Vagabond during the winter of my last year in high school, in the basement of Kinney's Shoe Store where I worked part time. Harold Taylor helped me carry the boat seven blocks for the shakedown. It floated.

Harold and I knew each other as HT and RT. He was a basketball player, and did not join in my longer trips across the river to explore the bayous and the densely wooded floodplain.

Down the hill from Rand Park, ten blocks east, the river road leads to Montrose. At the ferry landing two years ago, Paul Stoddart and I pooled our fortunes and paid $4.00 for a leaky jonboat. This was a substitute for the logs that we had assembled at the mouth of an Illinois bayou for the construction of a raft; they were washed away by flood waters. Because we would have to bail our way downstream to Mark Twain country, a decrepit jonboat seemed an appropriate way to reach the cave.

Keokuk's Main Street leads to Highway 218, the route to the DuPont explosives plant. To assure a margin of safety, there is a large tract of oak-hickory forest and the "powder works lake" outside a formidable security fence. When I was in high school, Kenny West, a cross-country runner, ran up and down the timbered hills around the lake as part of his summer training.

The Argyle Road, which curves around the DuPont land, leads to an old one-room school where I have often parked Dad's Model A. Following a fence line between a cornfield and woodland, I have reached the headwaters of one of my more productive creeks. The Indians left a lot of their ammunition in this land. I find arrow points and hide-scrapers on the gravel bars of the creek. Also, on the steep north slope of one ravine I find reindeer moss among the fallen oak leaves. My botany professor, John Sass, admires this leftover from the chill of glaciation. He believes that in Iowa the lichen is found no farther south than the ravines along the Des Moines River near Boone.

Inside KHS, things are much as they were five years ago when I walked across the stage. I reached for my diploma with my left hand, as directed, and George Davis shook my right. I assume he didn't expect me to return. Neither did I.

My homeroom at KHS is 209, next to the terraced lecture room where I learned a bit of physics from Mr. Willie Dodd and a bit of chemistry from Mr. J. C. Wright. Next in line is the laboratory for physics and chemistry. At the end of the hall, with windows toward 15th Street, is the biology lab, 119. It has no five-year memories for me. When I came to Keokuk as a junior, I was told not to enroll for biology; there would be too much overlap with the semester of zoology that I'd had across the river at Hamilton, Illinois.

The sixteen brass microscopes are in a cabinet to the left of the door. At the window side of my lab there is a long lead-lined sink, without a faucet. It may have been intended for storage of dissection specimens, impregnated with formaldehyde. I am sure there are better uses. With a lid of plywood, it will store empty fruit jars and provide a surface for such things as aquaria.

The boys' locker room and showers, down by the boilers, still smell of sweat and excitement. I wonder if anyone has topped Ed Cook's record in the shot-put.

Bordering 219 there is a small, dimly lighted museum room, walled with glass cases. They contain stuffed squirrels, badgers, orioles, and pickled rattlesnakes. I suspect there is more to be learned in Rand Park and along the creeks and rivers.

I found it strange, being there in familiar territory, yet in such a different role. How much of my former life would carry over into the teaching of biology and physiology?

R. L. Reid

Our superintendent of schools is Doctor Reid. The title is for some kind of medical degree in homeopathy. I have never met anyone who could remember when he was not superintendent of schools.

Dr. Reid is a small man with a round head. He welcomed us to our chosen task by announcing our slogan for the year: We, the people. It celebrates the adoption of the constitution 150 years ago, shortly before he became superintendent of schools. He kept holding up newspaper clippings to prove a point. Then he would forget what the clipping was about. After a pause, he would flourish the clipping above his round head and say, "WE, THE PEOPLE! Let us not forget that we are the PEOPLE!"

One of Dr. Reid's points concerned a rooster race. Brandishing a picture of roosters coming in on the home stretch, he exclaimed, "If we can teach roosters to race, oh what can we do for pupils who are not roosters, we the people!"

This prompted shop teacher Paul Megchelson to lean in my direction and whisper, "We could teach them to lay eggs."

Thirty Days of School

I shouldn't have needed an introduction to KHS. It is my alma mater. Yet when I walked through the halls, peering into empty classrooms, remembering, a new custodian came to my rescue. "If you are wanting to register, you're a day early. Sophomores tomorrow, juniors and seniors the day after."

That helped explain why principal George Davis had told me I should require students to call me Mister Trump. "If they call you Mister," he explained, "they are more likely to regard you as a teacher."

My first day of classes was disarming. The students were so quiet. In fact, they became quiet before I wanted them to. I wasn't ready.

If someone had thrown a paper wad, or if a fight had broken out, I would have known what to do. But this! Silence - it was unnerving.

Although I have forgotten just how I finally broke into this silence, I have some advice for beginners. Advice from a thirty-day veteran! Don't be misled by a quiet first day. Some of those quiet demons are quietly plotting.

If you are planning a tree identification field trip to Rand Park, don't tell the students to meet at the statue of Chief Keokuk. It is only 35 feet tall, overlooking the river from the highest point in the park. Some of the students will be unable to find it.

When you mention that the meeting time is 9:10 and someone asks, "AM or PM?"-act as though you didn't hear the question. Proceed with plans.

On the return from the field trip, if you glance back and find that the class is following single file, all in step, holding a pencil over the right shoulder in military fashion, don't stop; just be happy they're following.

When you are coating the tree labels with hot paraffin, don't leave the tray of paraffin on the table to harden. Returning for your afternoon class, you will find that some resourceful sophomore has sliced the paraffin into squares and passed it around the class. All but two girls in the front row will be vigorously chewing.

If you have ignored the paraffin advice, do not panic. Pass the tray around the class, explaining that students in the morning class chewed the same paraffin, but to your knowledge none of them has scarlet fever.

When someone immediately asks to go to the drinking fountain, say no.

Along with my two sections of biology (two double-periods each week for lab), I have a homeroom, one period of library supervision

(53 students), and two sections of boys' hygiene. How will classes of junior and senior boys compare with mixed classes of sophomores? I've been bolstering my scant knowledge of human physiology and health with lots of reading-back issues of Hygeia, and Dr. Clendening's daily column. Also, I am planning to tap some of the local physicians for advice.

When I told Dr. Holmes what I'm teaching, he asked, "Do you know how the human spleen functions?" As I started to say something about blood storage, he held up his hand and silenced me. "Richard," he said, "if you know how the spleen functions, you are the only person alive who does!"

Then Dr. Holmes asked if I had a stethoscope. Without waiting for my reply, he opened a drawer and pawed through some instruments. Bringing out several scopes, he examined them and handed me one. "You'll need a stethoscope!" he said. "A listening device for heart and chest sounds." He explained. I didn't explain that I knew what a stethoscope was.

My two hygiene classes are so different that they require different teachers. Oddly, the class that started with trouble has been more responsive. During the first week of school, four boys were absent for several days. They are all football players who according to rumors, were suspended for smoking on school property. While the four were absent, I made a seating chart. The day of their return, they arrived in class after work had begun and took seats in the very back row. I asked them to move forward. Their response was to go into a huddle, as though hearing the quarterback's decision.

Although distracted, I tried to go on with the lesson. Finally the big four came forward and took seats in the front row, close enough to reach up and touch my desk. Sitting there, they smiled as though they had just completed a touchdown pass against my team.

Next day, to my surprise, they occupied those same front seats.

This has been the more satisfying of the two hygiene classes. One day, as a change of pace, I read an excerpt from Don Marquis on the digestion of beans. They liked it. Their laughter was genuine.

In the other class, I soon concluded that I was making no headway with the beans. Halfway through a sentence, I put the essay aside and said we'd better get to work. Many of the students, I suspect, didn't notice the transition.

George Davis

Does Principal Davis remember that when I was a junior at KHS, I was summoned to the office following the opening of the baseball season? Keokuk was home base for a farm team-of the Cardinals, I believe. A goodly number of boys were absent in the afternoon, and I was one of them. Does my principal remember that I allowed him to give me the full treatment before I told him that I was a member of the Municipal Band, and that my absence had been arranged in advance?

Bertha Craig

Miss Craig has taught American history 42 times in Keokuk. Since checking me in and out of her class, she has added another president to the curriculum. The remodeling of the rotunda in Washington has not changed.

Last night we served together as chaperones for a dance. While we sat in the bleachers of the gym, watching the band tune up, I raised a question: "Why is it that while testing a microphone, people always say, 'Testing...one, two, three...testing?' They never finish a sentence." "A full sentence," replied Miss Craig, "calls for conviction. People don't care to commit themselves to that extent."

October River

Hank Elwood is just as exuberant on the river and in the woods as in class. I invited him to row across the Mississippi and search for a shellbark hickory on the Illinois floodplain. The water was higher than I'd expected, backed into the creek that comes down from the hills onto the forested floodplain. Having left the boat upstream on a sand bar, we crossed the creek on a fallen willow.

There was little daylight left, and we should not have gone so far. But the woods were so mysterious. We couldn't stop. I'd never been there in October. Although the trees are still green, there is a silent foreboding in the air, something you cannot see or hear. But you feel it. Perhaps it is not in the woods but in you, as it is in the shore birds that are beginning to fly south or in the muskrats that are hauling cattails into their dens.

Light was dim when we found the shellbark. It is a fine tree, at least sixteen inches in diameter, with great shreds of bark hanging loose. Looking against the sky, we could count the pinnately arranged leaflets, seven to a leaf, but we could not see any nuts.

Heading back toward the river, we stopped too often to look for pecans, and eventually we had to feel our way through the thickets and over the fallen logs. The uncertainty of it was wonderful. I was caught up with Hank's wild vitality, and I felt that I could go on crashing through the October woods until I fell exhausted.

Having misjudged the location of our willow-bridge, we waded the creek, and soon after putting our shoes back on, came to the open sky over the river and found our boat.

On the Illinois side, we could see the gray water edged by darker sand and willows. The shoreline was so different at night. So were Keokuk's hills in the distance. It seemed that we were rowing toward a shore that I had never seen before. It still seems that way.

Faculty Fatigue

A drama presented by the faculty is one of the annual George Davis fund-raising events. It's now over. The kids are still laughing. Staff members aren't. They are exhausted and irritable. The director, poor Miss Hutchins, deserves a long, long vacation.

The dress rehearsal showed only that we needed another week of practice. Several cast members were late. Mr. Davis needed frequent promptings. "Nothing to worry about," he said, "the play isn't until tomorrow night." Someone closed the curtains in the middle of the second act. The telephone failed to ring. (That's good, because there was no phone on stage to answer.) When James Wright (physics and chemistry) found that he was on the wrong side for his entry, he strode across stage in front of the action, disappeared in the wing, then made his entrance.

It's over, however, and I remembered most of my lines. Actually I feared that I might remember them too well. At one point, when asked if I was thirsty, the script called for, "I've been thirsty all my life!" In deference to Superintendent Reid, the script was changed; obedient, I said, "Yes."

I don't know about other members, but one bit of good came out of my fatigue. When rested and feeling good, I am too friendly. I laugh at little distractions if they are funny. They become big ones that aren't funny. In my present state, I am less tolerable. Today the result was that I straightened out several disturbers who badly needed straightening.

BIOLOGY FOR BEHINDERS

We're using an old textbook, Truman J. Moon's Biology for Beginners. About half of the book is in phylogenetic sequence. It progresses backward through the plant kingdom from seed plants to spore bearers. Then the animal sequence moves forward from protozoa to mammals. The second half of the book deals with man, his origins, physiology, and health. There must be a better sequence for studying the science of life. One of the books turned up in lost-and-found with a new title scribbled in: Biology for Behinders. No one claims it.

Teachers talk too much. I knew it when I was a student. Now I'm relearning it. I talk too much.

In my morning biology class we were discussing yesterday's test, clearing up misunderstandings. There was a restless run of whispering, and I had put up with it hoping that those who wanted help were getting it. When I tried to explain why a pollen grain is not comparable to an animal sperm, I surrendered. Without comment, I distributed paper and asked that books and notes be put away. A deadly silence followed. Then I distributed test questions, the ones we'd been discussing. My only comment as I collected the papers at the end of the period was that I would record the better score of the two.

This procedure wasn't necessary in my afternoon class.

I'm delighted with the response to a bulletin board display in the upper hall. With some pictures of wildlife and clippings on wise use of natural resources, I announced a meeting in Room 319 of students interested in conservation. It wasn't necessary to suggest the formation of a club. Schools are club crazy.

So a conservation club is the latest. The kids think it would be a good idea to have a hike and a chili supper and elect some officers, and maybe decide on some kind of project to work on. Do I really need another assignment?

One Semester Down

If I had kept the continuity of my log intact, it would contain bits of great satisfaction along with too much frustration and anxiety. Was George Davis right in saying that no one should begin teaching in his hometown? Should I have stayed with forestry?

I have done reasonably well with that first-name problem. I hadn't expected, however, that some of the students would be able to say Mister and make it sound like swearing.

Many of the kids appear really to be interested in biology. But many are followers, and easily led by astray by the outlaws. I once heard that if you place a hurdle in front of a line of sheep as they go out to pasture, then remove it after the lead sheep jumps, all of the remaining sheep will jump when they come to the place of the hurdle. The sophomores of KHS are like sheep!

They seldom challenge my authority. Yet a minor amount of chattering can cripple a discussion. It can cripple me, too.

One of the fine books I've read this semester is George Dorsey's Why We Behave Like Human Beings (like sheep?). It is a fusion of anatomy, physiology, and psychology, with bits of philosophy thrown in.

If I were looking at another person's life, I would say that after five years of higher education and several months of employment, he should know in what direction he is headed. With me, I don't. It may be Dorsey's fault. I tell myself if we understand why we behave like people, we should be able to decide why we don't behave that way. Could I accomplish more if I turned to human psychology?

I know! I'll teach a night-school course in human behavior. That will help me get my bearings. Next school year?

But why detour? In studying life, I am studying people. And so much more: the elegance of a minnow swimming upstream in a riffle, just fast enough to stay in one place, catching food that is swept toward it; the flight of geese, homing toward their northern nesting ground; the bursting of an acorn as the embryo becomes a tree.

Could I ever become a specialist?

Beryl D. drives an old Chevy that is generally equipped with a flat tire. Today he hadn't had time to read his lesson. The free air station was too far away.

Nancy R. is bored. She dates one of last year's graduates.

Jim G. and I have a little game going. We make the rules as we go along. I assigned seats a week ago, on Friday. By Monday, he had forgotten which seat was his and thought it best to take one in the back row. At my suggestion, he moved forward. On each of the following three days, he tried a different seat. Each time I merely motioned him forward. He came. Today Jim stopped in the doorway, and asked, "Where would you like me to sit today?"

On a quiz, Doris H. stated that crayfish eat "mostly little things but sometimes people." Wondering about the source of her information, I asked. "It's in our book," she replied. "It says they are sometimes cannibals."

Evelyn G. and Carita D. sit next to each other in the front row. About once a week they exchange whispers, looking apprehensively in my direction. I've not heard their voices.

Alec Smith on the crayfish: “It moves forward or backward or sideways or the other sideways. It seems to depends on where it wants to go. It doesn't turn around.”

Two Kinds of Antagonism

It is a part of my education, I suppose, teaching with Hale. At times, it is an uncomfortable part.

We are both in our first year out of Iowa State, both teaching biology. He has a section of physics and (this semester) a section of boys' hygiene.

A part of my antagonism is jealousy. Hale is better prepared to teach human physiology. He is a well-coordinated athlete. He can tumble and dive. He can go on stage and stir the students in a pep rally, leading off lustily in "Onward Keokuk," punctuating a cheer with a neat backflip. Although I doubt that I would use backflips as exclamation points, it would be nice to be so capable.

I envy Hale's greater freedom from discipline problems. He tells me that he doesn't care whether students like him, just so they respect him. I wish he could tell me that without smiling like an advisor.

Correcting my errors within hearing range of students is another matter. He could have let me go on pronouncing gila monster with a "g". When he heard of my report that an unusual Englishman could lower or raise his pulse rate at will, he could have asked about my source of information. Instead, he told his hygiene class that I was wrong. (Yes, Hale, I'm aware the heart is an involuntary muscle, but if the man can really "volunteer" his pulse rate, we'd better find an explanation.)

Our disagreement about muscle action has become a running controversy. In my hygiene class we were dealing with the muscles used in pull-ups, "chinning the bar." The biceps, we agreed, are a part of the muscle team that pulls you up. But what muscles do you use getting back down?

Having studied the antagonistic principle, some of the students said the triceps contract to pull you down. Others argued that the triceps aren't needed; gravity pulls you down.

When the students were unable to resolve the issue, I sided with gravity. I added that you use your biceps in lowering the body as well as in raising it. Without contracting the biceps, you would fall. By working the same team of muscles, mainly the biceps, pectorals, and latissimus dorsi, you let yourself down slowly.

A few days later, I learned that Hale had explained my "error" to his physics class, sketching lever systems to prove his point. Because a few of Hale's physics students are in my hygiene class, the disagreement continues.

When I told Hale that we should resolve the question ourselves and drop it, he agreed. He explained how muscles work in antagonistic pairs. I didn't "understand."

Such a View of Life

Today I watched students who were watching blood cells tumbling through capillaries in the webbing of a frog's foot. Such a view of life is all that William Harvey needed for proof of his theory that blood does not pulse back and forth through the vessels; it completes a circuit from arteries into veins.

My sophomores were viewing this amazing event as casually as they might the traffic of their classmates in and out of the rooms. An occasional face lighted with appreciation of the wonder beneath the lenses. But one group of a half dozen students, after waiting their turns, peered into the scope, nodded, and walked away without knowing that the frog had moved its foot out of the field of view.

On opening day last fall, I distributed filing cards, asking for information that I assumed would help me know my clients. For example: Why did you decide to take biology? Before collecting the cards I commented that biology is the study of life, and asked the students to write on the back of the card a brief definition of life. That little Irish girl, who I have come to know as the defiant one, wrote, "Life is what happens between wars." Today I heard her tell a friend that her brother got a letter from the draft board.

Degrees of Ignorance

Reader's Digest has scared my hygiene students about venereal disease. They want someone to tell them they are healthy. Knowing that the boys will feel more anonymous with an outsider, and knowing that I am handicapped by a degree of ignorance, I invite local doctors to come to my class.

The boys enjoy Doctor Cooper. His answers are quick and pointed. When one of the boys asked about the best way to keep from getting a girl in trouble, he replied, "The best way is to keep your pants on, zipped up." Another appeared to jolt the doctor: "How do you know if you are, well, how do you know if you are pregnant?" Hesitating briefly, Doctor Cooper said, "How do you know you are pregnant?" The boy nodded, and Cooper replied, "Maybe your girl friend can tell you." There are degrees of ignorance!

Catching Up With the Termite

One of the ironies of war: British ornithologists are predicting an increase in Europe's bird population in the next few years. Their reasoning: If people spend all their spare time shooting each other, they won't find time to shoot the birds.

It has been argued that battles are natural in the animal world. Song sparrows defend their nesting territories. Ant colonies raid and enslave neighboring colonies of related species. Some kinds of termites are so committed to battle that special castes of warriors have evolved, neatly designed for killing termites.

I wonder, will humankind evolve a caste especially adept at dropping bombs? You'd think that, having an edge on brain design, we might eventually discover that war doesn't pay.

If that time really arrives, perhaps we will also discover that it doesn't pay to shoot all of the birds.

Touch of Knowledge

I have wondered about this boy who wanders around the library and classroom. He touches things; magazines, books, bones. He touches them as though he is absorbing something of value through his fingers.

Perhaps he is. Today Jerry handed me a copy of Science News Letter and pointed to an item about vitamin B1. He thought I might have missed it. He waited as I read.

Yale botanists, I learned, have found rich stores of vitamin B1 in the buds and growing tips of a number of common trees: pine, elm, basswood, and others. When the investigators girdled basswood stems, cutting through the conducting tissue, they found high concentrations of B1 above the girdle, none below.

I said, "That's good, Jerry. Maybe instead of swallowing capsules, we should be chewing maple buds with the squirrels or browsing on the twigs of shrubs with the deer."

"Well," said Jerry, "Doctor Kritchel told us that when a dog eats grass, the grass scours out the dog's insides. Would those vitamins be in grass, too?"

Thanks, Jerry! When the searchers for knowledge of nutrition have lettered all of the vitamins and sifted out all of the traces of elements that regulate and lubricate, we may have a better understanding of animal behavior. You put your fingers on an interesting idea.

Birds, Ants, and People

This titmouse, taking a squash seed from my window sill, flying off to the safety of the poplar tree, looks as if he finds this earth a comfortable home.

I am comfortable too, watching him from behind glass. It is good, on a Saturday morning, just to sit, and perhaps lay down a book, and watch a bird.

The titmouse is back for another seed. I suppose it made many trips before I read that last sentence and laid the book aside. It was J. Arthur Thompson's Science Old and New.

"A common harvesting ant of South Europe," Thompson wrote, "collects seeds of clover-like plants, lets them begin to sprout so that the tough envelopes burst, exposes them in the sun so that the germination does not go too far, takes them back underground and chews them into dough, and finally makes this into little biscuits which are dried in the sun and stored for winter use."

An amazing idea, if ants have ideas. If they don't, it is still more amazing.

That is why, watching the hungry titmouse, I was thinking of that proverbial advice: "Go to the ant, thou sluggard, consider her ways and be wise..."

Then I was interrupted again. The radio was blaring something about a ship that was sinking off the coast of Italy. We live on a peaceful part of the earth, the titmouse and I. But we wish our European neighbors would spend more time looking at ants.

Faculty Came in Second

Another George Davis fund-raiser: during a lull in the basketball schedule, an assembly interrupted the day's work. It was announced that the men on the staff had challenged the varsity team. The assembly was called to learn whether the team accepted.

Captain Miller of the varsity replied that the team would accept the challenge provided Mr. Davis would appear dressed for action. Davis consented. He said he would wear a toga presented to him by the Counsel at Rome on an occasion when he had played an unusually impressive game before an international audience. All through the introduction of the staff, athletes there were spasms of unorganized cheering and jeering.

Then Miss Loeffler, the buxom math teacher, was summoned to lead yells for the staff. When she failed to get a response, there was a call for Miss Connor, then Miss Simmons, Miss Nancolas. Finally, the entire female staff was on stage, making feeble cheer gestures. There was more laughing and booing than cheering.

At game time that evening, the bleachers were filled. Within a few minutes, the varsity had scored 12 points. Wright, the varsity coach, tried frantically to direct some staff strategy; no one listened. One of our problems was that we never got the tip-off.

After a time-out, Davis came in as center. He isn't tall, but he compensates with considerable girth. Grether and I thought of hoisting Davis onto our shoulders for the jump at center. Each heaving on one leg, with three team members lifting from the rear, we succeeded. When the ball was thrown up, the varsity center was laughing and forgot to jump. Davis caught the ball. I'm not sure what happened next, but somehow the varsity scored a field goal before we could get Davis back on the floor.

Next we tried holding the Chiefs. We held them by the legs, by the arms, by the ears. This strategy often called for two of us holding one student, however, and this meant there was a loose Chief who could dribble off and score. All the while Wright kept pleading with us to stop clowning and play ball. (He was one of the five.)

During halftime ceremonies, cream puffs were served to the staff players. Believing that even if I could swallow mine, it wouldn't stay there, I motioned that I would share it with someone in the stands. Many eager hands went up. Removing it from the bag, I tossed it into the stands, and it was promptly divided into many small smears.

We rallied a bit in the third quarter. I grabbed a loose ball and was running down court with it under my arm when Bob Foster tackled me. For which I was awarded one free throw. And because the tackle had been "unnecessarily rough" I was allowed to rest awhile before shooting. To my amazement, this gave us our third point of the game.

During the fourth quarter, it appeared that the performance of the staff was deteriorating. A big sign appeared on the floor announcing a sit-down strike. We sat, panting. All except Hale. He did back-flips. During this rest period, one of the Chiefs made off with our sit-down sign and placed it over our basket rim. The ref declared that this was an impersonal foul worth one free throw. Because I had established a reputation for free throws, I was given the ball-with a warning that this did not include another rest period. The ball rolled around the rim so many times, I believe one of our players must have blown it in.

At game's end, the Chiefs had 51 points. Actually they scored 86, but they decided to donate 35 of them to us because of our sportsmanlike conduct. In addition, they had made two field goals in the wrong basket. We came in second with 41 points.

CONSERVATION

I wish there were a way for us to measure the runoff from the watershed above Nelson's Pond. The trees and shrubs planted by the Conservation Club and the contour plowing of the cornfields surely are reducing the flow of soil into the pond. If small check dams and good watershed management were more common, we'd have fewer floods. And our rivers would be cleaner.

TWO OF A KIND

Now that Hale and I are involved in a cooperative error, we should make better harmony.

We purchased a pair of guinea pigs for the lab. Knowing that students would ask which is which, we investigated. Hale took the one with the black rump, I the one with the brown shoulder.

"Mine's the female," said Hale.

"So is mine!" I replied.

We exchanged pigs. Same conclusions...General Biological Supply would hear about this.

Fortunately, there was a delay in getting a letter written. A couple of weeks later, on a Monday morning, we found a baby guinea pig in the cage.

Wright told us how we could explain this to our students. If one of the mamas wasn't a papa, there had been a stranger in the cage.

To Be an Explorer

When I was heading for a career in forestry, the lure of wilderness dominated my plans. After a year at Iowa State, however, I decided that people were more important than trees. And now, teaching biology, I find there can be a share of both.

Our home base is old Hotel Manning in Keosauqua. Across the big bend of the Des Moines River is the entrance to Iowa's largest state park, with more than two thousand acres that I must explore. Oh, it is well mapped. The early loggers who cut off the great oaks for railway ties surely knew all of the creeks and hills and meadows. I assume they knew where the vultures nested and where the woodchucks dug their dens. But for me this is all new. I am an explorer.

Perhaps the lure of wildness was guiding my unconscious mind in my first year of teaching when I stepped out of the registration line at the University of Iowa. From the talk in Keokuk, I had assumed that the proper way to get ahead in schoolwork was to enroll in Saturday classes and learn about school finance, guidance, and such. With J. C. Wright and several other teachers, I had gone to Iowa City to embark on a career.

As I stood there in line, shifting my weight from one foot to the other, I searched for reasons why I should take a course in public relations. There was so much I didn't know about earthworms and turtles and maple trees. Also, Lorene was in Ames; public relations!

Finally, I moved up beside Wright and told him I had changed my mind. I would hitch a ride to Ames. He shook his head in disbelief and said he hoped he'd see me Monday.

During the remainder of the school year, I spent half my time wondering whether I really had control of my classes. (That Cushing kid was a different species!) The other half of my time, I waited for June 17th.

After we were married, we floated down the Mississippi for six days in the skiff that I'd built during my senior year at Keokuk High. Then back to Iowa State for the second session of summer school. When I told my physiology teacher, Dr. Erma Smith, M.D., that I planned to begin work on a master's degree, she said, "You'll have ulcers!"

The following summer I was thrown off course by the offer of a fellowship in psychology. Although I spent too much of my time reading about bees, and not enough about educational statistics and the psychology of safe driving, the experience paved my way to a better understanding of the human species. And during the following school year I determined that I must spend more time out of doors.

That is why, on our way to Ames, I stopped in Des Moines at the offices of the State Conservation Commission. I wanted a summer job in wildlife work. V. W. Flickinger asked if I would be interested in the duties of a State Park Naturalist. I would receive $50 per month plus housing, and there would be time for a research project. I would. I am!

It would be fine to look out over a valley from a mountain peak that had never before been climbed. It would be fine to paddle a canoe along the shores of an unmapped lake. But for me, there is enough of the unknown in finding what ferns are here - in watching a nymph hump its back and crawl out of its skin to become a mayfly - in seeing the ways of the woodchuck.

This, too, is exploring.

Watching Saliva Work

As happens too often, I wasn't prepared for the lab. Too many interruptions before class. Too many items missing from the work stations.

This time I was lucky. As a warm-up for more exacting work on proteins with commercial enzymes, we were finding whether saliva really changes starch to sugar-in a test tube. It was relatively simple. I could give the student teams five minutes to decide on a suitable attack. In that time, I could distribute the test tubes and locate the dropper bottles of Benedict's solution.

Those five minutes might be the most important part of the investigation, at least for students who are already concerned about science. This exposes one of my prejudices. I dislike recipe-style lab work. Sure, students are more likely to finish on time if they follow step-by-step directions. And they are more likely to finish with the "right" answers. But . . .

On the chalkboard, I had written brief instructions:

Does your saliva convert starch to a reducing sugar in a test tube? Reporting your investigation, include these items - the question you are trying to answer - an account of what you did - what happened - why it happened.

I tried orally to emphasize a familiar admonition, "Those four items," I said, "are not an outline. Write your report in any pattern or style that you believe to be appropriate. Just be sure that each of those four kinds of information is included."

The chalkboard instructions included also a summary of the test for reducing sugars. After students had read the directions, I demonstrated the test, first on a dilute solution of sucrose. No color change. Then while the test tube materials were still hot, I added a few drops of glucose solution. In a few seconds, the color changed suddenly from blue to green. And when I swiped the tube through the Bunsen flame again, the green changed to yellow, then to orange. Now are there any questions? No questions. They knew what to do.

After remarkably brief conferences, they were spitting into test tubes, squirting blue liquid in, and waving them over burners. A few of the work teams made elaborate efforts to control all leaks in their procedure. I was delighted with Lori's dilemma. She left her team and told me despairingly, "No one will spit in ours!"

"Why don't you spit in it?" I asked.

"I've got fresh chewing gum," she replied. "I'm afraid it has sugar in it."

I told her to borrow saliva from someone who wasn't chewing. That delayed team C for a while, but others finished in record time, waiting for something to happen.

When the teams reported orally on their work, about half had recognized the need for controls. They found sugar, all right, and they knew where it came from. They had tested starch without saliva, and saliva without starch. One of the teams even tested the distilled water that they used for suspending the starch. No change in color. One of the teams that failed to include such tests was ready to argue. The bottle was labeled pure potato starch - no sense wasting time testing pure starch. And if there had been any sugar in Bill's mouth, he would have tasted it!

There were other problems, too. A few teams reported a negative test. "But we think we know why," said the speaker for team B. "We didn't give that enzyme, whatever it is, we didn't give it time to work." Another: "Maybe our saliva was too cold. Would that make a difference?"

One group decided that Linda was "too stingy with her spit." Jerry volunteered, "I gave them enough saliva, but I think they cooked it before they put the starch in."

In a 45-minute period, there wasn't enough time for reporting the tests to correct such mistakes. But I tried to cover with a final caution. "Tell in your written report exactly what you did. Tell also what you should have done."

I'm glad, however, that I don't have to define my own controls and really prove that this think-it-through approach is better than the cookbook route!

NECKTIE TROUBLE

Red-haired Darlene O. stopped in the lab today and asked if I remembered her. I'd forgotten the name, but I remembered the seat, just to the right of center, halfway back. Former students appear satisfied if you remembered where they sat.

Satisfied, Darlene fumbled with the thought that she shouldn't have dropped out as a sophomore, then told me what she remembered best about biology.

"Your tie was crooked in home room one morning," she explained. "It was still crooked in biology. We were having a lesson about something or other, and close to the end of the period, you asked if we had any questions. I held up my hand and said your tie was crooked and would you mind straightening it?"

"Wasn't it always crooked?" I interrupted.

"I guess so. Maybe," she replied, "but I mean the time I asked if you'd straighten it because it bothered me."

"Did I do it?"

"That's what I remember best about biology," Darlene continued. She flipped a strand of red hair from her eyes (I was glad - it bothered me). "You said you'd do it if it would help me learn biology, and everyone laughed. But then you got a comb out of your pocket and straightened your hair!"

Sorry, Darlene, but it's a relief to know why you dropped out.

COUNT TEN!

A week ago, I assessed Gerald S. a fine of 40 cents for scratching his name on his lab table. He grumbled and said he'd "see about it." A couple of days later I reminded him that he had not paid the fine. "What about Ray's table?" he asked. "You gonna fine Ray? He did the same as me."

Besides resenting his help in my responsibility, I didn't like his smile when he spoke to me. It was more like a leer. "Gerald," I said, "your fine is 40 cents. If it isn't paid by time I turn in my reports, you know that it will be doubled."

He said he would see about it.

Then today, as Gerald passed my desk, he said, without stopping or even turning to face me, "Have you collected from Ray?"

Now I have taken some pride in remaining cool under this kind of stress. I remember once asking my minister why he wanted me to be chairman of a committee that was plagued with controversy. He replied, "Because you have a long fuse." Well, when Gerald walked past, my fuse was ignited, but I told myself that it would burn slowly, and I wouldn't waste time and let the remainder of the class become involved. Maybe my spleen released some extra blood into my vessels, but not much. I wouldn't blow.

Gerald tossed his books on his table, leaned back in his chair and fixed his smiling leer in my direction. Scanning the room, checking for absentees, I could see that Gerald never wavered from his "I'll see about it" leer. But I didn't blow. Perhaps my breathing became deeper, but only a little. I wouldn't blow.

One of the girls in the second row, directly in front of me, was Barbara H. I was about to explain the day's work when I realized that Barbara was speaking, and I saw that her soft voice was aimed in my direction.

"Are you speaking to me?" I asked.

She nodded, adding a comment that I still did not hear. "If it's important," I replied, "you'd better say it out loud." And I suppose that a little more adrenalin flowed and my spleen responded. But why should I snarl at Barbara H., a member of the student council, a yearbook editor, a responsible science student? Well, maybe I didn't snarl, just glared.

Barbara reddened a bit. But she leaned toward me, cupped her hands around her mouth, and said in a forceful whisper, "I tried to tell you to count ten."

I don't remember much about the remainder of that period, we were studying blood pressure and perhaps I told about the function of the spleen. By the end of the period, however, I was able to smile. I walked to the doorway, and as Barbara passed, I whispered, "One!" And I turned away before Gerald S. arrived.

Knock on Trees

Why do I go around knocking on trees that have a woodpecker hole in a dead stub? Many of my ways have obscure origins. But I know why I knock on trees.

It goes back to my first year of teaching, on a delicious October day after the first frosts had labeled the landscapes with color and left the air tasting of autumn. One of my students had the good sense to say, "Why don't we go on a field trip? It's better outside!"

Well, we had completed our work on tree identification. And teachers knew, even then, that a field trip requires advance planning - a "felt need," a clear goal, follow-up applications, and all that. The only thing that fit was that I, too, felt a desperate need for being outside.

In ten minutes, we were strolling through Keokuk's Rand Park, without purpose. At one point, on the hillside below the pond, I stopped to wait for stragglers to catch up. I was standing by a red haw, thinking that a review of this specimen would help justify the outing. And I was staring up idly at a prominent woodpecker hole in a dead stub of the haw. As the laggards came forward, one of them noticed my interest in the hole. It must have been Bob F.

Bob glanced up at the hole, walked over and rapped on the tree trunk, asking, "Anybody home up there?"

That was one of Bob's better questions. He functioned mainly as the class clown. With his uninhibited wit and infectious grin, he found it easy to get the attention of his classmates. Knocking again on the tree, he asked, "Anyone home today?"

In response, a furry little face emerged from the hole and looked down at us. There was a cheer of delight, and with that, the flying squirrel scrambled out of its den. It raced to the top of the haw tree. There it poised momentarily, swaying from side to side. Then it launched into the air, gliding some thirty feet toward the trunk of a big red oak. Approaching the trunk, it lifted its tail, up-ending its body, so that it landed head up. And without loss of motion, it raced nimbly up the tree and disappeared in the browning leaves.

If we had been able to peek down into the nest before Bob knocked, we might have seen this little mammal asleep, curled up on its back or side. When it opened its eyes, we would have seen how big they were, typical of nocturnal mammals. It was probably resting on a bed of shredded bark, perhaps with bits of moss or lichens added.

And if we could have seen the squirrel perform in slow motion, its gliding would have been easier to interpret. The body skin extends in a loose fold between the front and rear legs. Leaping from a tree, the squirrel extends its legs outward. Air pressure bulges the loose skin upward, forming pockets of air, like the bulging of a paper kite in the wind. By leg movements, the squirrel can alter the slack in its skin on either side, veering to the left or right to dodge branches and control its landing.

There's an incredible account of the flying squirrel in the Journal of Mammology, May 1924. It is quoted in Seton's <u>Lives of Game Animals</u>. J.W. Stack, a forestry professor at what is now Michigan

State University, reported that eight of his students were trimming trees along a river when one of them cut a dead limb containing a flying squirrel nest. Discovering the nest, a student lifted it out, exposing four babies, almost hairless. While they were examining the young, the mother climbed a nearby tree, glided across the river, landed in a tree and appeared to search for a new home. After examining four large limbs, she glided back toward the original home, landing at the feet of the students. She immediately climbed up the pant leg of the student who was holding the nest, grasped one of her babies, climbed back down, then raced up a tree and launched out for the other side of the river. After depositing her baby in a hole, she returned and repeated her errand. In her four trips, her take-offs and landings were at the same points, though on one trip she first climbed the wrong student's leg. (Three out of four doesn't seem too bad a score!)

A couple of weeks later, Professor Stack returned to the site with a movie camera. When disturbed, the mother again moved her babies, now much bigger, to a new hole. This one was only 30 feet away. He measured the distance of the previous move and found that it was 86 feet.

Our own Rand Park squirrel responded to our knocking several times in the next few weeks. Then we saw it no more that fall. Flying squirrels do not hibernate, but they often move to larger winter homes where a number of them can huddle together, sharing their body heat.

During the remainder of my seven years in the Keokuk job, we had a fair number of field trips to Rand Park. At times, I suspect, the only need we felt was to knock on a red haw tree and see if anyone was at home up there.

Thank you, Bob, wherever you are!

BACKGROUND FOR CHEATING

If we want our kids to grow up honest, we should not pay them for good grades. And we shouldn't have too many kids in the first place.

That's implied in Lyle Johnson's report in Clearing House. He made it easy for 241 junior high pupils at Corvallis, Oregon to cheat on a test. He tabulated all of their answers without marking the papers. Then explaining that the scores would have no effect on their class records, he provided the answers and asked the pupils to check their own papers. Forty-six and one-half percent of the youngsters cheated by changing one or more of their answers.

Armed with a lot of data about each student, Johnson studied the relationship of various factors to cheating. Among those that appeared to have no significant relationship were the economic status of the parents, sex of the pupil, nationality, and whether the home was urban or rural.

Factors that appeared to encourage cheating were low IQ, having brothers or sisters (especially having older siblings), church membership, and rewarding the children for high grades. Johnson says that rewards are a strong incentive to cheat. He believes the reward system may be more prevalent among church families than among non-church families. This, he said, may explain the failure of "character-building" organizations in building character.

So, David Trump and any brothers or sisters who may follow, don't expect any dimes or dollars for good report cards.

Claws and Coronary Things

With plastic fingernails extending nearly an inch beyond the ends of her digits, she was having trouble fitting a cover glass onto a microscope slide. If she really wants to look predatory, why doesn't she fashion pointed extensions on her canine teeth?

Jody: "If the coronary things get plugged up, it can be very fatal." That's very true, Jody. And you might add that people who die of heart attack can be very dead.

How Many Microscopes?

I was called to the office for a telephone message. Vern Young, principal of the Ames Senior High, was on the line. He asked if I would come to Ames and teach biology.

I asked, "How many microscopes do you have?" He said, "Two." I said, "We have sixteen." He said, "We will get more."

When I put down the receiver and saw Bob Laughridge questioning me with his eyes, I told him about the message. Bob said, "Tell that man to stay away!"

Tuesday

I began again, this time at Ames High.

VIEW FROM AMES HIGH

While I was teaching in Keokuk, the high school in Ames moved west across Clark Street into a new building. The two microscopes and I are in the southeast corner, room 213. It is an all-purpose facility, suitable for discussions, yet reasonably equipped for laboratory work. The demonstration table, serving as an oversize podium, is on an elevated platform and is fitted with a sink, electrical outlets, and a gas spigot. There are fifteen worktables, each large enough for two students. Between my lab and the physical science laboratory, there is a narrow storage area with shelves and cabinets on both sides and a photographic darkroom at the west end. I like that west end!

Room 213 has ample windows south and east. They provide drab views across 5th Street of the telephone building, Pontiac sales, and the outlet store for Moore's Dairy. Across Clark is the old high school (now serving as a junior high school) where I survived my student teaching ordeal during my final quarter at Iowa State.

It is sad that a school should be so squeezed by pavement and commerce. Brookside Park, however, is only a ten-minute walk west, across a suspension bridge over Squaw Creek. The park proper is well groomed. Yet it has a number of floodplain tree species that I haven't seen in the residential area. One is the Kentucky coffee-bean tree, with several impressive specimens. And upstream along the creek is a less domesticated area, heavy with sandbar willows, honey locust, and prickly ash. Although I'll have to go farther for woodchucks, there must be squirrels and birds to fit the trees. Iowa State is only a half mile west of the park, beyond a pasture. This is easy bicycle distance, even from the house we are renting on East Sixth Street.

Leaving the Mississippi and the hills of southeast Iowa was painful. Still, this has attractions. And the move will make it easier to finish my woodchuck thesis.

There is a strong school spirit here - "Ames High aims high." Although I hear this mainly at pep rallies, I sense that the aim isn't restricted to athletics.

Vern Young says I am a good disciplinarian. Perhaps he is actually saying, "You'd better be a good disciplinarian." I am the warden for the final study hall of the day, and this is not the easiest time and place to enforce good order. Yesterday my charges grew restless during the final minutes of the period, and when an announcement came over the sound system, I was unable to get the message. Regaining control, I reminded the students that they may have missed important information. Then the bell sounded. Passing my desk, a girl remarked with obvious satisfaction, "Mr. Trump, it was about a faculty meeting right after school."

Harry Knight is a fine fly-grabber. He can catch them barehanded as well as some of us can with a net. I remember one of his demonstrations during an animal ecology field trip. Someone asked Dr. Knight how he did it. "Don't grab where the insect was" he replied. "Grab where it is going."

The sophomore counselor told me that Louise plans to become a biology teacher. I hope she cures her lisp. It became obvious when she was struggling to pronounce photosynthesis. In a couple of attempts, she got as far as "photothin..." then recognized that it was wrong and started again. Thinking this was only a temporary failure of the tongue, I said, "You mean photothinthethis?" And she replied, "Yeth." She'd better plan on teaching zoology, never mentioning where animals get their food.

From Thad's report on a canoe trip on the Skunk River: "It appeared that the wildlife was very wary because all that we saw was out of sight. One time we saw a fish jump two feet out of the water twice in a row, but I don't know if it was one fish twice or two fish once."

From the middle row: "You know those bristly little caterpillars that you see on walnut trees. They aren't very big, but they're awful bristly. What do they turn into?" I was glad Gordon spoke up. I didn't know the answer. He said, "Little tooth brushes."

Feast of Bones

Even as a sophomore, Howard has found a way of leaving his mark on Ames High School. His grade sheet will probably show a list of indifferent C's. And his counselor's remarks in the permanent record folder will say, "Howard does not appear to work at his full potential."

Today, however, he contributed to the museum in 213. We were discussing the economy of the natural world, how molecules are recycled among living things. There's not much waste of substance. As an example, I took an old bone from the cabinet, a bone I had picked up in the woods.

"Look at this," I said. "Bones don't lie around long in the wild before some rodent comes along, sharpens its teeth, and gets at those valuable minerals. I'll pass it around. Look at the evidence." Handing the bone to the front row people, I added, "If any of you feel the need of calcium or phosphorus, help yourselves."

As the bone moved from table to table, we continued the discussion. And when we were getting to the way fungi clean up on some of the leftovers, I noticed that the bone had reached Howard.

He was holding it in both paws, gnawing.

Students around him were watching. I paused in the middle of a thought and stared. That caused others to turn to see what was happening. Some of the shorter students stood up for a better view. Undisturbed, Howard continued to gnaw. When satisfied, he examined his accomplishment, drew a pencil line around the scratches that he had inflicted, and passed the bone along.

It will remain among my treasures.

Why Hens Don't Need Teeth

Howard's incisors remind me of the steel ball bearings that spent varying periods in chickens' gizzards. To demonstrate that lack of teeth does not deprive seed-eating birds of a way to grind hard food, I encouraged each of five chickens (destined for our kitchen) to swallow a steel bearing. No problem. It rolled right down. Recovering the bearings at intervals-one day, three, six, nine, and twelve-I have a permanent display in a glass tube. The one-day treatment merely dulled the surface of the bearing. Three days in a gizzard left shallow pits and grooves. The roughening is greater on the six-day bearing, and the one from nine days is the most deeply pitted. The twelve-day bearing, has strangely symmetrical grooves, like the longitude lines on a globe of the earth.

I tried scarring a steel bearing with a machinist's file; it wasn't easy! I doubt that even Howard's molars could match the work of gravel in the muscular gizzard of a chicken.

(Much later, I learned that more than 200 years ago Reaumur had tested the strength of a turkey gizzard. He used a sheet-iron tube that could be dented only by a pressure of 80 pounds. The turkey flattened it in 24 hours. And Spallanzani fed a turkey a dozen steel needles and found that they were ground to bits in 36 hours.)

My Round of Golf

When you tell a joke on yourself, people should laugh with you, not at you.

Taking over my duties at Ames High, I discovered that in addition to teaching biology, I was coach of the golf team. I was open-minded about it. Never having played golf, how could I be otherwise? For the practice sessions at the Homewood course, I rode my bike out, counted the team members, who had arrived ahead of me in a car, then got lost in the rough where a pair of Cooper's hawks was nesting in a white oak.

Next morning the students would come to my room and post their scores on a bulletin board chart. They told me this was how the former coach decided who played in the next meet. The golfers were not very clear in explaining how this worked. It involved a cumulative score, with a little extra weight on the more recent practices. But, since the boys seemed to know who would play, I didn't interfere.

On one occasion, the team members seemed rather disappointed in my performance. At the invitational meet on the Fairview course in Des Moines, a controversy erupted over some kind of ruling. On the way home Bruce C. remarked, "Mr. Trump, you should have protested!" Intent on driving the Chevy that I had borrowed for the trip, I said that things like that generally work themselves out if you are patient.

At season's end, Ames High had done reasonably well. We came in second in the state meet. (I believe this means that only one team among the Iowa finalists had a lower score.)

Now in the awards assembly, the coach is expected to say a few kind sentences about team spirit and strategy. Knowing that word had leaked around concerning my coaching techniques, I thought it best to explain how little I knew at the beginning of the season and how much I had learned from the boys. I started with the nesting behavior of Cooper's hawks. By watching Hal W., I had learned how to grip a putter, and was finding the skill helpful in hoeing my victory garden. And near the end of the season I had found that a birdie and an eagle are not the same; one is better than the other.

Several students applauded, and teachers smiled at me in the halls. But for a week or more I heard derisive little comments about birdies and eagles, from low-minded students who seemed to believe they had discovered and exposed my deficiencies.

I suspect Principal Vern Young has spindled a memo concerning a golf coach for next year. Oh well, whoever - he probably won't even learn the alarm call of the Cooper's hawk.

Anson's Contribution

For nearly a week Anson L. hasn't asked when we're going to dissect something. I suppose it's because of what happened a week ago during the fifth period.

Anson is the long boy with light hair. He sits in the third row, his feet protruding from beneath his lab table into the second row. He works after school for a local pharmacist, unpacking shipments, delivering prescriptions, and cleaning the windows. Occasionally, he sells cough drops, aspirin, and such. He says you learn a lot about people's troubles, working in a drug store. And he believes he'd do alright at dissection.

His question generally came while I was checking attendance. "We going to cut up something today?" he would ask, even when he knew we had an exam scheduled. Rather than looking at me for the answer, he would turn toward Helen B. By the time he turned, Helen would be contorting her face by contracting the muscles that raised her upper lip and expanded her nostrils. I wondered whether he would have asked about dissection if there had been no girls in the class.

Well, last Wednesday I wanted a drop of blood for some preliminary observations of cell types. Anson pulled his legs into the proper aisle and came expectantly when I motioned him forward. "We going to dissect?" he asked, looking into the sink where we kept a frog and a salamander.

I turned to the class. "We need some blood for today's work," I said. "Anson, you know, is interested in surgery, and he has offered to be the donor. You can go ahead with your notes until we have the blood." This, of course, was just a way of telling the students to observe the blood-letting.

Anson was looking at the animals down in the sink. When I started taking equipment from a box, his feet shifted; and his eyes, but not his head, turned so he could see what I was doing. First, I took out the 200 ml beaker, holding it up to the light, and blowing off a bit of lint. Then, a bottle of alcohol and a large roll of absorbent cotton. Next, a scalpel, which I examined deliberately. A white towel, and, finally, a can of diethyl ether, with the label toward Anson. He shifted again, glanced at the class, and quickly back into the sink.

Arranging the equipment with elaborate care, I paused again to inspect the edge of the scalpel. Nodding with satisfaction, I pushed a lab stool into position and motioned for Anson to be seated. He was gripping the corner of the table with his left hand. His knuckles were white.

"Roll your sleeve up," I whispered. "No, your left. You're right-handed, aren't you?"

During this preparation, the class had been unexpectedly quiet. I had assumed that the boys would be free with advice and encouragement for Anson. They appeared to be too curious to let themselves go. And, with his staring into the sink and his furtive glances at the scalpel and beaker, Anson, too, was taking the ordeal in silence. When I doused a large wad of cotton with alcohol and swabbed Anson's arm from the elbow to the fingers, there were a few nervous titters from some of the girls.

Everything in order now, I quickly took two microscope slides from the drawer, held Anson's finger, and pricked it with a lancet that had been hidden by the edge of the towel. After touching a slide against the red drop that oozed out, and completing the smear with the other slide, I turned to the class.

"Anson," I announced, "has contributed to the cause of science!"

Anson bowed, and there was a burst of applause. He then turned and looked at the equipment on the demonstration table, shrugged, and walked to his seat. When Helen leaned across to look at his finger, which he was holding at an angle from the others, he shrugged again.

Canned Biology

It's like instant coffee: you just add water. With canned biology, you add sophomores. When I explained how it works, some of the teachers nodded approval. I'm not sure about the reactions of the other panel members. They were seated behind me.

My argument was that biology teachers occasionally need a canned lesson, one they can lift out of a file or a cabinet, one that won't require supervision. "Free" time: glorious minutes for recording grades or planning tomorrow.

This was a Tri-County Institute. A team of speakers from the DPI and from centers of learning like Drake, Iowa State, Iowa U. and Ames High spent a day in each of three county seats in southwest Iowa. We were charged with the awesome responsibility of inspiring the county's public school teachers.

As an example of canning education, I told about my indoor "field trip" for the study of insects. It includes about 20 displays of insect specimens, together with fragments of information about the life histories and ecology, the kinds of comments a teacher might make when conducting a genuine field trip. Some of the specimens are in Riker mounts, some pinned on a cork stopper that is fastened on the inner surface of a jar lid. One is a prepared microscope slide of a dog flea. Another is a jar of frass left by the termites that invaded a supply of stationery that was stored in a Des Moines basement. There is a plastic-covered cabbage leaf, ragged from the work of cabbage loopers, with one of the loopers in a small vial.

All of these items are kept together in one box. I can drag it out on short notice when bad weather interferes with field work or when a bad checking account balance has interfered with sleep. Students are instructed to move around the room, beginning at any point, to take notes on what they learn. Next day, or next week, I can use a short quiz, allowing use of their notes. Which of the insects in the displays have sucking mouth parts? Which ones have an incomplete metamorphosis?

This particular canned lesson has a special advantage in the study of insects. In a single hour, students can see a diversity of insects that would be hiding somewhere during a real field trip. The main advantage, however, is that at the end of the day, I feel I've been paid for not teaching biology!

In Harms Way

When I arrived at Ames High John Harms was teaching physics and chemistry. He was famous for his one line response to an amusing question. . His wife had given birth to twins, a boy and a girl. A friend stopped him on the street and asked if the twins were identical. "No," said John, "they have different fathers."

John Harms left Ames High for the U.S. Navy soon after I arrived. Plump and sedentary, he was not a choice candidate for the physical rigors of indoctrination. He knew it. Yet he yielded to Ray Smalling's urgings. Ray, before enlisting, a member of the Ames High coaching staff, had already completed the indoctrination. "John," he wrote, "you can do it. Come on in!"

After a few weeks in the Navy, John wrote what I suspect is the briefest thank-you note on record:

Dear Ray,
You sonofabitch.
Sincerely,
John.

Interruption

I interrupt my teaching for my part of a war.

Aw, Come On, You Guys!

While I was with the U.S. Navy, I received a questionnaire from a hometown group concerned with the welfare of their men. Among other items, I was asked how my experiences were contributing to my vocation. I replied that I had learned to shave in 30 seconds. I've since recalled a priceless experience at the indoctrination school at Princeton. Every Tuesday and Thursday at 1300 we funneled into a warm auditorium to watch an educational film. And sleep. One particularly exciting film concerned the chain of command by which you could request overseas duty on a mine sweeper.

The sleeping phase seemed to annoy the young officer in charge.

After turning on the lights at the end of the film and gently awakening us, he would explain the importance of educational films. They might save our lives. When this failed, he tried turning on the lights during the film. This, too, had little effect on our sleeping habits. Finally, he begged us to cooperate. In a tearful whine he said, "Aw, come on you guys. You're going to get me in trouble!"

Afterward I heard one of us remark! "If we have trouble sleeping next time, let's pretend we're asleep."

What did I learn? If begging doesn't work, don't show educational films at 1300 on Tuesdays and Thursdays.

Demons of the Sand

Ant lions earn their living by trapping small insects, injecting a paralyzing drug, then sucking out the nutritious juices. My introduction to these larvae of lacewing flies came in the sands beneath the coconut palms and breadfruit trees along Manzanillo Bay in Panama. I was there in 1943 with the U.S. Navy, awaiting sea duty. My ship, one of the Liberty freighters, was headed for destinations in the Pacific, and I was scheduled to take command of the Armed Guard crew. During the five-week delay, surrounded by palm trees, land crabs, and exotic birds, I began to think that my part of the war was a rather gentle thing.

Under one of the breadfruit trees, however, I watched a struggle that was not so gentle. A large black ant had blundered into the crater of an ant lion. The lion quickly seized the ant and pulled it partly beneath the dust. But the ant jerked loose. When it started up the side of the crater, the dust provided poor traction, and to make matters worse, the lion immediately started throwing soil particles and dust, starting miniature landslides. Several times the ant skidded to the bottom of the crater. Each time the ant lion grabbed but lost its grip. Finally, the ant reached the rim of the crater and escaped.

At other craters, smaller insects were not so fortunate. They were quickly pulled below. Hours later, I would find the victim, little more than its exoskeleton remaining, outside the pit.

Because ant lions excavate their ant traps at night, I brought several of them into the Officer's Quarters in a can of dust. They spent an hour or so plowing through the dust in reverse gear, just below the surface, making a maze of furrows. This appeared to be a matter of locating suitable locations for their traps.

Having settled on a site, an ant lion backs around in a circle about four centimeters in diameter, throwing sand by quick backward jerks of its head. Gradually decreasing the size of the circle, it eventually hides at the bottom of a conical crater, with only its tong-like mandibles visible.

Wondering how the ant lion keeps its pit clean, I returned the skeleton of a small beetle to the bottom of the cone. The owner promptly tossed it out with a vigorous thrust of the head. The same thing happened when I placed a small flat stone in a crater. But, when confronted by a larger stone, somewhat more than half the size of the larva, the strategy changed. After failing to throw the stone over the edge, the lion backed out of the crater, pushing the stone. Leaving it beyond the edge of the pit, the lion turned around and backed into its insect trap, then buried itself again.

Watching the excavation of pits in cans of dust, I was impressed by the accuracy of the spacing. With three larvae in a coffee can, the pits marked the corners of an equilateral triangle. Allowing reasonable distance to the side of the can, the pits were spaced as far apart as possible.

What would happen, I wondered, if the habitat were overcrowded? When this happens in a muskrat range, the weaker individuals are driven into unfavorable environment where they generally fall prey to minks or foxes. Quail respond in a similar way, keeping the population down to the carrying-capacity of the land. On a limited scale, I tried overstocking my cans with ant lions. It appeared that the principle applied. The population was disrupted because the insects threw dust in each other's pits. Some of the displaced larvae plowed through their neighbors' traps. Since ant lions are known to be cannibalistic, this might thin out the population in a crowded natural situation.

The Captain Scares Easy

Now I know why superstitions die hard. I learned on a dark night on the Pacific. I was aboard a rusty Liberty ship in the middle of the ocean when a tired bird flew into our forward gun tub.

We stood peering over the gray water. The civilian captain pulled at his fingers, cracking the joints. "I hope the men don't hear about it," he said.

They had warned us about the men who live long at sea on merchant ships. We were to appear tolerant. It would help keep relations between the gun crew and the deck hands on an even keel.

But I was still unprepared for that night when the bird came aboard. I had scoffed at spooks too long. Oh, as a boy I had been afraid of mad dogs and kidnappers and tractors that looked like German tanks, and raspy insects that sounded like rattlesnakes. Things like that, but not of goblins. Strange sounds, but not silence.

So, I laughed at the captain that night on the Pacific.

I had prodded the old bos'n, too, when we were watching the wake of the ship one afternoon and our conversation turned to signs of luck.

"A dog," explained the bos'n, his voice droning with the ventilation fans, "a dog ain't smart on a ship. But you take a cat - a cat is smart. Did I tell you about the cat the mess boys brought on the tanker we was sailin' out of Newport? Well, when we's halfway down the coast this cat went back under some gear on the fantail and had herself a bunch of kittens. And she'd come around of a morning lookin' for flying fish that's stranded on deck."

"That's real luck," I goaded. "How many kittens?"

"Now you listen, lieutenant!" He took his foot off the rail and faced me accusingly. "You got something to learn about cats. We's taking this tanker in to Pensacola, then on to Caracas. I was there, lieutenant, an' I saw it myself. One day before we got to Pensacola, this old cat came along carryin' one of her kittens. She had it by the skin of its back, an' she took it forward and climbed it right into number three lifeboat. Then she went back aft and got another kitten. Ever one of those kittens, she took all of 'em right into number three!" The bos'n paused, as though it should be unnecessary to finish the story. When I looked up, he added, "Everybody was scared the ship was goin' down. Cats are smart on ships."

"Well," I pressed, after another pause, "did it?"

"Don't know," he replied. "I suppose so. I didn't take a chance. Got myself replaced at Pensacola."

I nodded approvingly, but laughed inside, the way they told us at the Armed Guard School in Gulfport.

I guess I didn't hide my feelings as well with the captain. He had sent a seaman to my quarters to say there was a bird in the gun tub forward, and I ought to know about it.

Feeling my way along the damply salt-glazed rail, then between the trucks and Jeeps that were lashed to the deck, I reached the forward lookout. He nudged something dark on the deck of the gun tub. "It's a bird," he said. "I got my helmet over it."

While I took the binoculars and scanned the sea, the gunner found a wooden crate, which we inverted over the bird and weighted with a round of dummy ammunition.

On my way aft, I stopped at the bridge where the captain was chewing out the third mate for being off course. I thanked the captain for the message.

"What'd you do with the bird?" he asked.

"We couldn't see much," I explained. We decided to keep it till morning."

We stood staring toward invisible horizons.

"I hope the men don't hear about it." He turned on the third mate. "Don't you tell anyone, third - you listening to me? Birds are bad luck when they die on a ship."

There were more knuckle sounds from the captain. He glanced aft, then faced me again. "That's right, lieutenant! It's no good forcing an animal off a ship, dead or alive."

He looked over the mate's shoulder, pushed the wheel over a fraction, then back. "I'll give you an instance. I was on the run to India with a tanker right after Pearl, and we had this dog that kept fouling the decks. So when we tied up at Suez, some damn sailor took this mutt ashore and left him. Five days later we were torpedoed and took to the boats."

He looked at me, underlining his words as though they were coming directly from Merchant Marine law. "You don't force an animal to stay aboard, lieutenant. You don't force one off the ship. It's bad luck all the way!"

He pushed the third mate aside and worried over the compass. "It's damn bad luck, lieutenant."

Breaking in pale foam against the port bow, water stretched away into darkness. The bird was on Navy property. The Armed Guard was my command. The bird wouldn't die.

I returned to my quarters muttering that the third mate was right. "The captain scares easy."

The watch had changed, and I was asleep when someone knocked at my door. I recognized the second mate. "Lieutenant, the captain sure wants to see you on the bridge."

Straining back to reality, I stepped into my shoes and groped up the ladder. The captain was pacing from one side of the bridge to the other, thumping his fists together.

"Listen," he gasped. "We got trouble. A radio distress. There's a ship three degrees south got a torpedo across the bow. A near miss. I thought you'd want to warn your men."

Suddenly it was colder. The ocean bigger. I buttoned my jacket, shivering as I felt my way forward between the trucks. I was walking in the wrong direction. I knew that. I should have returned to my cabin and used the battle phone...three degrees, about fifty miles to a degree at this latitude. It could have been a porpoise. I thought through my remembered instruction. A sub close in, open fire forward and try to ram. (Try to ram, with six tons of bombs in the holds?) Farther out, turn tail and fire aft. (Let's be sure to spot it far out!) Add more armor-piercing projectiles to the ready boxes. Remind the men in dim light you see best out of the corner of your eye. And don't let them go below without polaroid goggles.

With the battle phone, I could have alerted the watch in an instant. But I was walking forward. And as I climbed to the gun tub I could see the gunner on watch, silhouetted against the gray sky. "Things okay?" I asked. "You're wide awake?"

The gunner said yes, and I recognized the voice as Stewart's. "Keep a sharp watch," I cautioned. "A ship three degrees south is in trouble."

I felt warmer. Stewart was a good man.

Yet I stood there, as though there was more to say. How long, I don't know. I stared at the dark water and the dim sky, thoughts floating around my mind, making silent words come to my lips. No point in keeping the bird longer. We'll let it fly if it wants to. What's a bird got to do with submarines? If the captain comes telling you what to do, let him know this is Navy territory. Tell him you need my permission. Why should we keep the bird till morning? Even in daylight I won't be able to identify it. This is a matter of principle. Birds don't decide where subs will strike. Which way is south?

I tried to find a familiar constellation, but the stars were dimming out in the gray of predawn.

I started down the ladder, then stopped and said, "Pass the word to your relief, Stewart."

"What word, lieutenant?"

"Oh, I guess I hadn't mentioned it. Don't let anyone bother that bird. I want to see it in the light. Then we'll let it go, if it wants to go."

Tuesday – Continued

After the war, I began again!

Now I have learned to sit on a thumb tack and pretend that I am comfortable.

Return to Duty

When I returned to civilian life in early April, 1946, Adams asked if I would take over in the biology room right away. There were six weeks remaining in the term, and an interim teacher had "already finished the book." With an opportunity for fieldwork on woodchucks and cottontails, I declined. Knowing the teacher's desperation, however, I helped her expand some of the expended topics with magazine articles, special reports, and fieldwork.

It Started with Stink Bombs

I didn't ask to teach another subject. I only agreed that I was better qualified than Miss Nowlin. She is the home economics teacher. During the first semester, she was assigned the overflow section of chemistry.

It was while I was giving the semester exam in one of my biology classes that principal Herb Adams came by and explained the emergency.

"The students have learned how to make stink bombs," he said, "They're putting them in the juniors' lockers. You know we can't have that." I nodded, and he continued, "I've checked your credentials. You are qualified."

I should have told him, yes, I had some work on bombs in the U.S. Navy.

But I didn't really learn that much. So I said, "Legally, yes, I am qualified. Let's see, I took my last course in chemistry twelve years ago. I can do it if I have some time for reviewing. The students have a whole semester's head start on me."

"I understand," Herb replied. "you'll have all weekend. And we'll relieve you from the study hall duty."

Herb Adams has a versatile face. He can frown and smile at the same time. While his brow creases with troubled furrows, the corners of his mouth can turn up toward his ears. "Actually," he concluded, and the corners of his mouth dropped a few degrees, "you may want to start with chapter one. I understand that some of the students haven't learned much besides the formula for stink bombs." Herb was funniest when he was deadly serious, when the corners of his mouth dropped. But I didn't laugh.

I started Monday. That was five weeks ago. I have reviewed my assignment as a way of rationalizing my being only a half step ahead of the students. (It took one week for me to learn how they are getting the ferrous sulfide for the bombs.) This review explains also why I am so considerate of the students' welfare. I watch closely for signs of stress.

Well, this is Friday, the day of shortened classes when we insert an extra period, between the second and third, for hobbies. I sponsor the photography group. When our speaker, Jon Morgan, arrived, I explained the situation. "My chemistry class meets here right after your talk. When the kids come in and see that you are still here, they'll want you to stay. They are expecting a test. It's all right with me if you can spare the time."

I was right. Dick C., with his good strong voice, was the spokesman. "Photography is all about chemistry, isn't it? Stay and tell us about it. Mr. Trump won't care." "I'd have to have a glass of milk," said Jon, glancing at the clock. "I have to feed my ulcer every hour."

Dick rushed from the room and was soon back from the cafeteria with two bottles of milk. Jon drank, then talked. He explained how he had never liked chemistry either and how he could never make an outline or write a speech and how a photographic emulsion is mainly silver halides. The period half gone, the questions were tapering off, and Jon said he thought it was time for him to get back to work. That brought another flurry of questions. Esther said she wasn't sure she understood what he had said about the fixer, and Ted reminded him that he hadn't explained how you make pictures in color. And so on.

When the bell rang, Jon thanked Dick for the milk, and the students left nudging each other with satisfaction. What they didn't know was that I had not yet finished making out the test.

~~~

In my third class, wisecracks erupt so readily from Gordon S. and Dick D. that when they fail to take up an obvious cue, I feel something has gone wrong. Sometimes I am tempted to pause a second, then inject, "Did I hear somebody say...?"

But I am learning to keep my balance, and perhaps to hide my disappointment.

Today a student teacher from Iowa State was expounding the accomplishments of Louis Pasteur. "No one could figure out why the wine was spoiling," she said. "But Pasteur's wife had given him a microscope for his birthday. So he looked through the microscope at some of the wine, and what do you think he saw?"

There she paused. She paused long enough for any self-respecting sophomore to reply, "Tiny green elephants!" Or maybe, "Drunken snakes!"

But there was no answer. I strained bravely, biting my lower lip. Presently the teacher answered her own question with a bland description of "tiny little microorganisms."

Where were you, Gordon? Where were you, Dick?

BRIEF GREETING

When Vivian walked into the chemistry room, the usual pre-class chatter stopped abruptly. She had registered several days late. She wore a black sweater. I suppose that the eyes of some of the girls narrowed in response to a possible invasion of their territory. And I know that some of the boys stared, while others looked quickly down at their books and then stole little glances.

If I had known Dave P. better, I might have guessed what he would say when a new girl came to sit beside him. But Dave, too, was almost a stranger.

Checking my class list, I found that everyone except Dave had a lab partner, boy with boy, girl with girl. They were seated in roughly the same pattern at the classroom tables. That's how it was when Vivian
~~~

came in and stood beside me with her admittance record. Dave was alone at a table for two.

I debated momentarily. No, it would be embarrassing to be given a choice. "We do our lab work in teams of two," I told her. "Fortunately there's one place left."

Dave pretended not to notice her approach. "Dave," I said, "this is Vivian. She's from New York. And Vivian, this is Dave. You two will work together in the lab."

I stalled there just long enough to see Dave look up at Vivian. And now that I know him better, I realize that his greeting was the most natural thing for Dave to say. Not "Hi!" or "Welcome to the rat-race."

He needed only one word, and he said it for all to hear. Raising both hands above his head, he announced, "Female."

Ron Easter

During the spring of my senior year at Iowa State, I was job-hunting, and one of my interviews was in the little town of Irwin in western Iowa. While awaiting my turn to confer with the Board of Education, I met another applicant for a vacancy in the high school, a Drake University graduate, Ronald Easter. He later signed a contract.

Now Ron has joined the staff at Ames High, taking over the work of John Harms. Our labs are separated by a narrow storeroom, with doors opening in both directions. Yesterday the physics classes had been working on gyroscopes, and I heard Ron telling his class about the difference between a magician and a teacher. "Magicians try to make simple things look complicated. Teachers try to make complicated things look simple."

Occupational Hazard

I soon recognized that one of my frustrations is a common affliction among teachers. During a class discussion, I watched Amy and Joan, who were visiting intently. They were living in another world, and it would serve them properly to be humiliated. Amy first.

"Amy," I said, "give us your opinion."

Without hesitating, she said, "Well, I think chordate is a better term. There are some, like sharks, that really don't have bones."

How did Amy know we were discussing chordates? How humiliating!

And I think of Mark D. He was in one of my early physiology classes, an intelligent son of a policeman, frittering the semester away, wasting my time as well as his. One day after class I asked, "Mark, when are you going to start living constructively?"

"Right after I graduate," he replied. And he did, doggone it, he did!

A Poorer Planet

Marsh hawks eat more mice than quail. The red fox eats more gophers than pheasants. As the argument goes, predators are our friends because they eat our enemies. By thinning the population of mice and gophers, however, the predators provide more habitat in which our enemies can live and reproduce.

We say that because the hawk and the fox find it easier to catch the slower quail and the weaker pheasant, they upgrade the fitness of our friends. But don't predators upgrade the fitness of the mice and gophers? Of course.

Friend and enemy are poor ecological words; the effects of predation cannot be assessed by a simple formula. But I think we can say with conviction that the loss of species is making the earth more fragile, more susceptible to chaos. Predators are particularly vulnerable; our planet would be poorer without them.

The Boil

Ron Easter and I were conferring at the front of the chemistry lab when Bob M. approached.

Bob: I want to check on my unknown.

RT: I gave you the answer. What do you want to check?

Bob: I want to know how there could have been any phosphates when I didn't get a precipitate.

RT: We're busy right now, Bob. We're selecting a new text. I'll be with you later.

Bob: I just want you to know I think your records must be wrong.

Ron: Now we know.

RT: Better go on with your work, Bob.

Bob: I haven't anything to do.

RT: Do you have something filtering?

Bob: No, I'm boiling the filtrate.

Ron: Well, go boil with it!

As part of a chemistry quiz, I asked for the proper way of disposing of various waste items such as matchsticks, strong acids, and broken glassware. Inspired by the practicality of the question, I added "unidentified chemistry notebook" to the list. Here is John F's answer: "Tear it into tiny pieces and sprinkle them on the countryside from an airplane. Then spray the land with a good disinfectant, and don't grow crops on the land for 50 years."

Why We Need Two Eyes, Two Ears

When I ask a class whether there is really an advantage in having two eyes instead of one, I am often told that it's nice to have a spare. Right, you might lose one eye when hit by a bean blower. Any other advantage? Yes, you have a wider field of vision with two. And if they were spaced on opposite sides of the head, more like a robin's, you would have an even wider field.

Those two ideas come easily. The contribution of binocular vision in distance judgment is less obvious. If you focus on a near object, as in reading, your eyes turn so they both point toward the same part of the book. If you are watching a TV screen across the room, your eyes are aligned for more nearly parallel lines of sight. Your brain interprets these angles, telling you the distance. The rangefinder of a camera uses the same principle.

To underscore this advantage of binocular vision, I have asked students to hold a pencil or pen in each hand, and with arms nearly extended and one eye closed, touch the two points together. They, discover that it is easier with both eyes.

The advantage of two ears? I generally start with an observation rather than the question. A blindfolded student sits at the demonstration table, facing the class. I stand close behind the student and snap my fingers in various locations. The student tries to point to the source of the sound. When the source is off center, nearer one ear than the other, there is seldom an error. When the snapping is equally spaced with respect to the two ears, above, below, in front, behind, there are many errors. It is curious to see the subject point in different directions when the snapping remains in one place.

Students who are observing find the errors hilarious, and the person who laughs loudest is the next "volunteer."

After several of these observations I ask, "How do your ears detect the direction of sound?" Less observant students have failed to notice that the errors come when the sound is directly aligned with the head. Those who are more alert not only see this alignment, they also conclude that there are two possible explanations of the ears' ability to locate a sound. The sound is louder in one ear than the other. And it reaches one ear sooner than the other.

As a spinoff, we have tested the response of a blindfolded student who is wearing a stethoscope. She or he is to point toward the apparent source of the sound, which comes from a very light tapping on the diaphragm of the scope. What the student doesn't know is that there are two diaphragms, connected separately to the ear pieces by rubber tubes. A helper stands in the middle of the room, holding the two diaphragms. When I tap on one, then the other, the subject points appropriately to the right or left.

Next, the helper quietly carries one diaphragm forward to one side of the room. I carry the other diaphragm to the opposite side. The tubes, however, are crossed. Now the subject invariably points to the wrong side of the room.

I seldom hear of any biology lesson taken out of 111. Yesterday, however, one of the librarians asked what my students were doing, standing behind someone, snapping their fingers. "I asked one of the boys. He was sitting there with his eyes closed, and his girl friend was standing behind him snapping. He didn't open his eyes. He just said they were doing biology. It looked innocent enough, but I'd like to know what's going on."

Snake in the Hand

Last day of school! Report cards were to be passed out at 2:30. Not before. So I put them in a small drawer of my desk, then went to the office to use the sound system for announcing the next yearbook staff. Reading John R's name, I visualized John roaming about the room, exchanging opinions on non-school matters with his friends. So, after stating John's assignment, I added in an undertone, "Sit down, John!" Returning to my room to distribute the cards, I found John wandering calmly out of the storeroom. I suspect he had not even heard his name mentioned.

The remainder of the students seemed even more anxious for the report cards. No wonder. When I opened the drawer where I had stowed them, there was our big bullsnake coiled over the cards. To complete the year, I wrapped the snake around my forearm, its head at my wrist, and held the cards in that hand. When the bell rang, the students came by in alphabetical sequence and took their cards off the stack. Ellen looked at me and pleaded, "If you'll give it to me with your other hand, Mr. Trump, I'll promise not to take biology again next year."

The Day of the Earthworm

It was the day of the earthworm. Working in teams of two, the students were to find how the five hearts are connected with the remainder of the circulation system. George and Mark, however, had gone beyond the call of duty in sectioning their specimen. The parts were a hopeless jumble.

"Bury it," I said. "It will make good humus."

I should have been more explicit, telling them to wrap it in the paper towel and get on with the written work. While I was helping Ellen and Martha locate the dorsal vessel, I was aware that George and Mark were folding a sheet of paper to form a small box, roughly the shape of a worm. And while I was urging Joe to take the dissecting needle out of his lab partner, George and Mark excavated a shallow trench in the planting bed, beneath the big geranium. I assume they used their scalpels as shovels.

After the class had left, I found that the boys had erected a tiny grave marker that may justify dissection in the curriculum. It read, "REST IN PIECES."

Points

I hadn't recognized how much I resent "points" until my student teacher from Iowa State made a suggestion today. Rich was briefing me on his plans. He would prepare a work sheet that students could turn in "for extra points." Although I could feel myself bristling, I couldn't say why. I told him we'd talk about it after school.

While scoring a couple of subjective test items, one on adaptations of the human spine and another on the attachment of the flight muscles of insects, my thoughts kept straying to points. When Rich came in, I said that if he had suggested a worksheet for those who wanted to learn more about mitosis, I would have approved on the spot. Offering extra points is too much like selling a grade.

It is easy to find buyers, and I guess that's why I resent points. They become a medium of exchange. Students are led to believe they are getting a bargain.

Track Record

Despite my objection to the way athletics dominates the thinking of many educators, I still dream of competing. Last night I dreamed of leading in a long sprint. I was comfortably ahead but was afraid I might have lost count. Maybe there was another lap.

Then I saw a tape stretched across the track, timers poised with stop-watches. At last, I'm going to win. I'll touch the tape. I did. I felt the slight pressure across my chest, and heard a timer call out 53.9. Then my muscles shifted into neutral, and momentum carried me on down the track.

My dreams, I suppose, are an effort to improve on my track record. As a senior in high school, I was awarded a varsity letter, but the situation was irregular. If I had been an asthma victim, I might have been called to the stage during the awards ceremony. As it was,

Coach Wright happened to meet me in the hallway after the assembly. He was carrying a leftover letter, Keokuk's purple K, with wings. Handing it to me, he said, "You made a good effort, Richard. You deserve this." Now if I'd just had asthma! Everett B . had it. He was a half-miler. He made a good effort. He came wheezing down the track, trailing by thirty yards. If you could hear Everett coming, you knew you weren't winning. If I'd been a half-miler, I would have heard him coming.

As a final item during the assembly, Everett was called to the stage. As he came forward, his wheezing was muffled by vigorous applause. Wright explained that every team needs competition. Everett competed. He ran in every time trial. He was handed a purple K-with wings.

I competed, too, but without asthma.

Three years later, I was a sophomore at Iowa State. With my weight lifting, rowing, and running, I was a half inch taller and three pounds heavier. That brought me up to five feet, ten and one half inches, and a hundred and thirty-four pounds. Along with the exercise as a book-runner for the library, I was ready for the intramural track and field meet. After I had cleared the bar in the final round of the high jump, I was shaking the sawdust off when a tall, tall man appeared beside me. His coach's whistle hung from his neck at about the level of my chin. I looked up and saw that he was wearing glasses, round lenses with thin dark rims. He looked like Coach Bob Simpson.

Simpson asked my name. And when I told him, he said, "Trump, I'm going to make a high jumper out of you!"

At KHS the high jump had been my only hope for a letter. But Anson J. always went an inch higher. He seemed to glide over the bar, his legs bent more than was common in the western roll. I had changed to the roll from the old scissors style that I learned as a freshman at Kahoka. I had trouble keeping my head from going over before my legs. Diving disqualified.

My strongest threat in the high jump came during an invitational meet on our home field. When the heavy thunderstorm hit, Johnson and one other jumper had cleared the bar. Five of us had failed. Because of the storm, I was in a five-way tie for third place. So that is the sort of talent Simpson discovered. He would make a high jumper out of me.

Because Oscar Hatch Hawley was trying to make a clarinetist out of me, I was late for track practice. When I arrived at the lower level of the men's gym, a string of runners were conditioning on the track. An assistant coach spotted me and told me to fall into that line of runners. As the end of the line came by, I fell in.

I jogged along unaware that the last man was expected to sprint past the line of men. That put another man in the rear, and he was expected to start passing. That way you alternated a fast jog and a sprint. Completing the first lap, I was still the behinder, and the assistant called out, "Pass 'em, pass 'em!" I did. And when the next man moved up to pass me, I stepped up my pace. Coming around again, I lengthened my stride.

Next time around, both Simpson and the assistant stepped out on the track. Pointing, they said, "Let 'em pass!" I did. And after one more lap, they stopped the line. That tall man with round lenses was looking down at me. He said, "Son, what's your name?" I told him again. "Trump," he said, "I'm going to make a runner out of you!" Simpson probably didn't know that I had run only half as many laps as the remainder of the squad. Neither did he know about the clarinet. Or the library. He produced a number of champions. One of them broke Simpson's own record, as a hurdler, I believe. But he didn't accomplish all that he set out to do. If he had, maybe I wouldn't dream of running.

First Day Fears

Time must have dimmed the fears of my first days alone with a class. At least I cannot remember terror of the kind Linda S. is facing. Though she is still in the observing stage as a student teacher, I suggested that she take over one of my afternoon classes and let me work on the bees. She had observed the lesson in two earlier sections, and I assumed she was ready to solo.

The lesson was a review in which pairs of students came forward and helped their classmates with questions assigned yesterday. For example, in which situation would you expect competition to be keener - between two closely related species or between two remotely related species? Linda was familiar with the ideas, and she had seen that the plan worked smoothly.

"Think it over," I said. "Let me know after lunch. This will give you the advantage of being fully in charge without a critic in the room." And I assured her that the work with the bees could wait. When I returned from lunch, I found Linda alone, standing at the front of the classroom. She was staring at the empty seats. "What's your decision?" I asked.

She looked toward the empty seats again. Then, almost in a whisper, she said, "You'd better go out to the bees before I let myself say no!"

It Didn't Help!

Al R., one of my eighth-period sophomores, has no class during seventh period. Finding that my lab is vacant during that time, he began coming to class early. This progressed until he now makes 111 his home base for the free period.

Al apparently enjoys watering the plants. And perhaps, because I've referred to his green thumb, he has carried buckets of soil into the greenhouse and trimmed the geraniums. He seldom initiates any conversation.

This afternoon, however, Al shocked me into listening more carefully. He was examining the willow shoots that were rooting in pots on the windowsill.

"Their leaves are so little," he said. "Why is that?" "Maybe they don't have enough light," I replied.

Al then opened the curtain, which had been closed during the previous period for a film. He stood there for a moment examining the willows, then called to me as I was about to leave the room. "It didn't help!"

WHERE IS MY HEART?

I knew only her name. She was one of the fourteen girls in the third period of my day. She came quietly into the lab, chatted with Susan at the second table in row C, and completed her reports on schedule.

When I asked for another way of determining the age of the ailanthus stem, no one volunteered. But when I confronted her directly, she said maybe the spaces between terminal bud scars would tell the age. Then last Friday her place at the second table was vacant. I recorded her absence.

Today I learned that she is dead. On Thursday night, she had called from her bedroom and asked her mother which side her heart was on. Then there was a sound of gunshot.

I wonder if too many others, like me, knew only her name.

BEANS IN THE DARK

The cartoon that Rick brought in today shows a frustrated student working on a home project. She is inspecting two potted plants. One, scrawny and wilted, is labeled, "Rich Topsoil." The other, a sturdy specimen with deep green leaves, is labeled, "Eroded Clay." The distraught student is saying to her dad, "Maybe I could just switch the labels around."

Rick thought the cartoon might comfort some of the students who were perplexed by our greenhouse experiment last week. We planted ten rows of beans across the bed of soil, and just as the seedlings were emerging, we covered alternate rows with light-proof boxes.

Because most of the students knew that plants require light, there was no great excitement about the outcome. A week later I suggested that during a study session the students go out in small groups. In our tiny greenhouse five is a mob! They were to peek under one of the boxes. Allen's comment was typical. "They grow faster in the dark? That IS weird!" One voice, however, was saying, "I told you so, stupid. We learned that in the seventh grade."

We were not yet involved with plant physiology, and I avoided any explanation. I said we were collecting some data that would be useful in a couple of weeks.

On the tenth day after covering the plants, students returned to the greenhouse in teams of two. Working on an assigned pair of rows, they cut off at soil level one plant from a lighted row, one from a covered row. They measured both plants from soil level to the tip of the terminal bud, then placed them in separate trays, one for all of the lighted plants, another for all of the covered plants.

While the teams were recording the measurements on a chart, some still thought there was something wrong. The average height for the exposed plants was 127 mm. For those in the dark it was 193 mm. Perhaps we should shift the labels!

But there was another surprise. After counting to verify that we had equal numbers of plants for the two groups, we dried them overnight in a low oven and checked the dry mass. The results: a total of 44 grams for 123 plants in the light, only 37 grams for an equal number in the dark.

Now we are using the data: In every class a number of students recorded in their notes that the leaves of the lighted plants were larger, the stems thicker. Of course? Beans need light to manufacture food. The covered seedlings had to get along with the food that was stored in the seeds. Still, this would not explain the greater height of the covered plants. After all, those in the light had the seed food plus the food produced by their own leaves.

Why? "When you read about auxins in preparation for tomorrow, watch for answers. And when you complete your records of the experiment, don't switch the labels."

POCKET GOPHER

Don H. brought in a male pocket gopher. It had probably strayed too far from its hole, mate-hunting. It cowered in the corner of a small wooden box with a screen lid, its upper incisors overlapping its lips. Having a living gopher presented a problem. I consulted the class. "We don't have enough time for everything today. Which shall we do, work on the outcome of dihybrid crosses, or take the gopher out and watch it dig?"

One smart-alec thought we shouldn't interrupt the study of genetics.

So we took the gopher out on the lawn and formed a tight ring of feet and legs. We agreed that since gophers are vegetarians, there would be no screaming if it ran toward someone's feet. It didn't. When I pulled the screen loose and stepped back into the circle of students, the gopher moved out of the cage and started scooping through the turf with its powerful front legs. They are short and heavy. Pushing the soil beneath its body, it was soon half out of sight. Once it appeared to kick the loose soil back with its rear legs.

In an amazingly short time (I wish we had timed it), the gopher's tail had disappeared. Then we inverted the box over the hole and pushed a trowel down, prying the rodent back into the cage.

"Well," I said, "that's that. I guess we still have time for genetics."

To which that same smart-alec replied, "Next time, Don, catch something that doesn't dig so darned fast!"

Food Chain

I am on a predatory bird binge, and a few sophomores share my enthusiasm. Last fall, Darrel L. brought in a marsh hawk that had been shot and hung on a fence north of town. To find what it had been eating, we opened its stomach and recovered the skulls of four field mice, the white-footed species. And a month ago Jimmy K. found a freshly-shot red-tail. From its stomach, we recovered three of the less digestible parts of a small bird: feet, bill, and gizzard. Matching with specimens in the Iowa State Museum, we learned that these parts had belonged to a slate-colored junco.

What had the junco been eating? Opening the tiny gizzard, we found an assortment of weed seeds, mostly from foxtail grasses. Someone suggested that we plant the seeds. We did, and now there are several foxtail seedlings growing in a pot at the south window. If the mice don't find them, maybe they will produce seeds that we can pass on to a hungry junco. Next in line-a red-tailed hawk.

This is what I learned from one of my cynical superiors in the Navy. If you know you are going to be out-of-bounds, carry a clipboard with a sheaf of papers on it. When you see someone looking your way, start flipping through the papers and don't look up.

Gruesome Question

I am compiling possible answers to the gruesome question students ask when they return to school following an absence. "Did I miss anything important?"

Here is a start:

1. No, we've been waiting to see if you recover.

2. Which do you want first, good news or bad? If the student says bad, I'll reply. "We had a test. They say it was easy. The good news is that you don't have to take that particular test. I have a different one ready for you."

If he or she wants the good news first, I'll say, "We had a review test. They say it's a stinker. I'll give you the bad news after I've checked your answers."

3. Yes, you missed anything important.

Neat Dream

Several boys were around a table at the back of the lab when I came in after lunch. They failed to notice my entry, and I heard Eddie tell of a "real neat dream."

It concerned a biology teacher and a woodchuck den: "Old Trump wanted to know the temperature down in one of those dens in the courtyard. He crawled down in the hole till just his feet were showing, and he got stuck. We were watching, and we wondered should we get a hold of his heels and pull him out. We were still trying to decide when I woke up. Maybe he's still down there!"

That boy has missed something. There is a four-foot length of conduit extending from the surface into the den. We can lower a thermometer on a string, leave it for a while, then pull it up for a reading. The next time we check the den temperature, I must send Eddie. When he returns to report, I'll tell the class that I had a neat dream last night. I dreamed that one of my students crawled down a woodchuck den to check the temperature and got stuck, with just his feet sticking out. We were all standing there trying to decide whether to pull him out when I woke up.

What good thing?

In another effort to upgrade Ames High, the administrators have asked each of us to decide, "What good thing would you like to see happen in this school?" I will see how this looks on paper and perhaps submit it as my reply:

I would like to see all teachers act on the conviction that whatever else students learn, a vital lesson is that if they do not work, they fail. If we deny the students this lesson, we fail. And I include myself among those who need to act on this conviction.

Wednesday

Now that Sputnik is up there, even the president of the board is in favor of science.

Perhaps he is afraid the Soviets will come and socialize his funeral business.

WHAT WE DON'T KNOW

On her way out of class, Doris stopped at my desk long enough to tell me that my tests "always show what we don't know. Can't you give a test that shows what we do know?" Then there is Cheryl's final thought about plants: "Without plants we would starve to death, die of suffocation, and expire from a combination of deficiency diseases. In brief, life would hardly be worth living."

TODAY IS A PART OF YOUR FUTURE

Checking your registration card, Jim, I noticed a change. On the line for vocational plans, you have erased something and written, "Foxhole, Korea."

At first, I thought this was another of your jokes. Then it occurred to me that I haven't heard you laugh in more than a month. Your grade point has dropped. And there's that trouble over training rules.

Does this have anything to do with Foxhole, Korea?

At a recent staff meeting, someone mentioned the international turmoil, the confused morale among our students, and the two dropouts joining the Marines. Before we turned to registration procedures, a lot of good advice was wasted because you weren't there.

Eight of us in that meeting were only a few years out of the foxholes and gunboats and bombers of another war. We were all teaching before we put on uniforms. And we all came back.

Now the odds are heavy, Jim, that you'll see military service, too, and that you'll come back.

One of your first assignments in service will be to go back to school. I think you'd be smart to finish high school now, and to do a good job of it. Lives are lost through the blunders of fellows who can't follow directions and who can't think straight. That's why they'll send you to school. Whether you are trained to grease machine guns or translate code, the way you can serve your country best right now is to finish high school. Military people agree on that. Finish school, and wait your turn for a uniform.

But while you are waiting, Jim, don't forget your real plan, the one you erased. Keep going with your piano lessons. Start college, too. Toting a gun may seem a long way from the training you'll need as a musician; but real musicians are real people. Their music is an expression of worthwhile living. With your intelligence, you can come out in a few years with a future that's even brighter because of your experience.

Remember, though, that today is a part of your future, too.

The Skull

I am puzzled why I should like Jeff. We all know he is a rascal. Pilfering lockers, forging passes, foul-mouthing his teachers. As his math instructor remarked in the lunch room, "We'll see that boy's picture in the post office some day, full front view and profile."

Yet, I enjoy having Jeff in class. Certainly, I enjoy him more than Grif, who calls me sir. I wonder if I like Jeff because he thinks I like him. It could work that way. Or, do I like him because I defended him last fall when he was about to have a fight with Grif?

Jeff is a tall boy. Tall and loosely hinged. His bones are maturing faster than either his muscles or his morals, and I have wondered if there is a kink in his chromosomes.

Last fall when our students were about to join the university crowd in a boil of antiwar protests, our staff discussed ways of cooling the fires. We decided to set a day when most of the classwork would give way to talk about "issues," whatever teachers and students agreed on. Let the kids speak out.

It apparently worked well in some classes, particularly with the seniors. In my classes, there was only one volunteer. It was Jeff, with a speech that lasted less than a minute.

When he came to me at the first of the period asking for issue time, I assumed he was working up a blast against Ames High for being a concentration camp. (One of the incidents that prompted our plan was the planting of a long tangle of barbed wire atop the outside wall of the unfinished west hallway. A crude sign identified the wall as Stalag 18).

I agreed to Jeff's request without asking about his topic. As he turned to go to his seat, however, I called him back. I said, "Jeff, watch your language!" He nodded and grinned.

After the class had assembled, I reviewed the agreement. This was a time for listening as well as for speaking. "We have only one volunteer," I said. "Are there others?" No one responded, and I motioned to Jeff. When he stood up, two dozen pairs of eyes turned, and there was a restrained titter from the corner of the room. As he came forward in the center aisle, I moved toward his empty seat, thinking it would be good to join the students. To allow for two-way traffic, we both veered to the side, and Jeff bumped Jan's table, knocking a book to the floor. There was another small laugh, stifled in the middle.

Stopping and turning in front of the demonstration table, he looked at the floor momentarily, then faced his classmates. His boney Adam's apple moved up, then down, as he swallowed.

"It's about the war in Viet Nam," he said.

Having started, he seemed to gain courage. "I think it's a stupid war, and we ought to get out of there. And if you are drafted, go to jail. Or if you don't want to go to jail, go to Canada."

That's when Grif laughed.

It was a crude laugh, unsympathetic. Jeff glared. "You trying to say something, Griffith? If you try, it'll be stupid. You think you know how to wrestle. You're a fake at that, too!"

Grif stood up and started forward. "It won't take me long to show you who's a fake!"

Seated at Jeff's table, I was able to move quickly between the two boys. Motioning Grif to be seated, I asked Jeff if he had finished his speech. He shrugged.

"Are there any comments or questions?" I asked. "This is your chance to say what's on your mind."

Perhaps because attention had been diverted to personalities, there was no response. I thanked Jeff and said I hoped the Viet Nam thing would be settled before he'd be called. He returned to his seat, glaring at Griffith.

In the months that followed, Jeff continued to earn his low C, despite unexcused absences. There were a couple of disciplinary conferences in the office. The vice-principal summed it up when he said, "Jeff is a renegade."

Yet we continued to have a kind of unspoken understanding, Jeff and I. Though I could not explain it, I felt it.

Ames High was quiet today-the strange quiet that you can hear on the day after the last day. A locker door would open in the hall when someone returned for gym clothes or a last load of books. I had turned in my grade records and was working on the attendance report when Jeff walked in and handed me a small box. Opening it, I found a smaller box, a safety match box. Sliding the cover off, I discovered a tiny mammal skull. I tilted the box for a better view. "A lot of teeth," I said. "Either a bat or a shrew. Which is it?"

"It's a bat. I put it on an anthill, the way you said. The ants sure cleaned the bones."

"I'm glad to have it, Jeff. It is our only bat skull. I'm grateful!"

"I thought you might want it."

And Jeff was gone.

A Little Older

When Bernice B. was in my class, she told me of her plan to skip her senior year at Ames High. "Two years of high school are enough," she said. "I want to go to college."

I agreed that she was bright enough, but wouldn't it be helpful to allow another year for maturing?

On several occasions during the remainder of the year, we exchanged jibes about her plan. I would remind her how great it would be as a senior at Ames High. She would reply that it would be greater to have a head start in science.

Today, during fall registration, I met Miss Black on my way out of the office. "Bernice," I said, "you look a little older this year."

"Mr. Trump," she replied, "you look a little older, too."

They Learned How to Work

I asked C. N. Brown how he accounts for the illustrious leadership of his three sons. Norman was president of Ames High students in '47. Wayne followed four years later. And now it is Keith. All three are leaders outside of school, too. When I confronted their dad with my question, he replied, "They have a good mother."

"Yes," I agreed, "they have a good father, too. But what is it that makes you and your wife good parents?"

Brown looked from the barn to the house, then back to the barn. "Well," he said, "all I know is that the boys learned how to work."

Unprincipaled High School

An unprincipaled school? Why not? The idea surfaced one day at lunch when a few of us were fussing about a series of class interruptions that came from administrative actions. We will establish our own school.

It will be staffed only by teachers who teach. A secretary or two, but no administrators. Each of us will make her or his own rules. This occurs to a degree even in well-regulated schools. In ours, however, students will quickly learn that consistency is unreal. They will learn to assess the vagaries of their superiors.

We recognize the possibility that one of us might begin to show qualities of leadership. For example, by suggesting that we agree on uniform ways of enforcing promptness. If this should occur, the offending teacher will be shunned for 30 days. And if the same person suggests that we meet to discuss behavioral objectives, the sentence will be extended to 30 years.

Those were our thoughts.

Now that I am a day older, I realize that leadership is inevitable. It will emerge, just as a weed seed eventually germinates. A teacher will innocently suggest that we pool our resources and buy a pound of coffee. The same person will drop a hint that absenteeism is increasing, and that there will be less vandalism if we lock our doors. Such initiative will lead to a lighter teaching load. With increased enrollment, the parking area will become congested, and the leader will appoint a committee. Another cut in the leader's teaching assignment will follow.

We will begin to consult the leader about the choice of time for field work, and whether students should be required to complete History A before taking History B. The leader will then require a 21 percent higher salary and a private secretary. Soon, when the secretary answers the telephone she or he will say, "May I ask who is calling?"

Finally, the leader will have an office, remote from any classroom. The office will bear the leader's name, with a distinctive title. It will be Principal.

Then those of us who teach will rejoice. There will be someone to blame for irregularities.

There are several possible prospects for the possible shunning that I propose for an unprincipaled high school. First to come to mind is Roger Spratt. He is young, tall, with exceeding energy. He is brash enough to step forward, sensible enough to get our attention. These qualities seem to override an obvious deficiency: He doesn't like coffee.

Leave the Birds and the Bees to the Bees and the Birds

I've been reviewing what goes on among birds. Perhaps I'll not tell my students what I have learned. They've been studying bees, and that's enough to undo their minds for awhile. The virgin queen going off on a mating orgy, producing a family with a dozen different fathers, the males not having any fathers at all, and not doing any work around the house, having only one contact with a queen and exploding in midair while performing their function, the workers driving the males out of the house into the autumn frost.

Why did humans ever start telling their kids about bees? It would be simpler and safer to start with people. And as for the birds, here are a few exhibits:

A. Courtship is an elaborate ritual in many bird species. They bow to each other, rub bills, sing, give gifts, and more. As a conservative example, the song sparrow: Staking out nesting territory in early spring, the male picks fights with other males that try to move in on his domain. He sings a lot and builds up to frantic dashes toward his prospective mate. This often ends in collisions. Ornithologists call it pouncing.

We don't need any additional pouncing in the halls of Ames High.

B. Some courtship antics are downright reckless. Grebes try to walk on water, and they do it. Some of the swifts indulge in high aerial chases, almost out of human sight. They copulate in midair, during a spectacular free fall, but they regain control before crashing. And bower birds collect trinkets for decorating their nesting bowers. Although home decorating seems acceptable, these birds expend a lot of energy stealing each other's ornaments.

C. It's a problem in certain species that the sexes look too much alike. Before courtship can become practical, there must be agreement about who's who. Some of the penguins have solved this by a simple inquiry. The male (who presumably knows that he is a male) brings a pebble and drops it at the feet of another penguin. If he has made a mistake, the other male starts a fight. If there's no fight, the pebble-dropper knows he is in the right room.

Now this may have a little merit. The boys are turning to girlish hair, and the girls are wearing trousers. But dropping rocks in the hall? No, I don't think it should be encouraged.

D. In many species, the male courts by displaying some distinctive feature of his species. The peacock turns his back on the pea hen and fans out his fantastic tail. The sage grouse inflates his esophagus so that it bulges like a balloon below his chin. Lacking these adornments, the male black-crowned night heron has a red iris. When special effects seem appropriate, he bulges his eyes part way out of their sockets, exposing more of that red tissue.

This wouldn't do at all. Too many sophomores wear contact lenses.

E. The male house wren is sometimes cited for being a bit promiscuous. Another way of putting it: the pair bonds may endure for only a single brood. In fact, after the eggs have hatched, the female may leave the male in charge while she flies off to find another mate and start another family.

Let's not go quite that far in liberating the girls.

F. Then there are the phalaropes, ocean-going shore birds that are experimenting with an alternative life style. Among other species, you might say in the establishment, the gaudy male takes the lead in courtship. He hangs around home, luring predators away from the nest. The drab female is less aggressive in courtship, and less conspicuous when tending her nestlings.

In the phalaropes, however, the female is the gaudy one. She is the aggressive suitor. But after the eggs are laid, guess who takes over?

This is fair enough for phalaropes. But when women have demonstrated that they can be more colorful than men and still do a better job of feeding the family...well, I believe I'll drop this one.

G. As though the white stork doesn't have enough problems of its own, it has been assigned the responsibility of distributing human babies. This started with the notion that the nesting of storks on your roof brought good luck. That was in the days when good luck included having a lot of babies. Now people are prevalent enough without having storks clapping their bills and building their nurseries on top of ours. No, I don't believe I'll add the stork to our curriculum.

They Just Sat

It would have been a fine relief today if they had thrown their books or marched out of the room in protest. I could have found a way of coping.

They just sat.

They reminded me of sleepy chickens, whose extra eyelid can slide down over the cornea, shutting out most of the environment. They just sat.

It would have improved my day if they had all fallen asleep and slumped onto the lab tables. Then I could have walked quietly away to find what odors of early March had come to the prairie. Now the Army's Corps of Engineers is proposing a dam across the Skunk River between Ames and Story City. The resulting lake would be considerable. It would sell a lot of outboard motors and water skis. And it would destroy a lot of good farm land. The flood plain forest would become a grand graveyard of leafless trees, their roots buried under tons of silt. Another treatment of symptoms instead of disease!

Preview of Suffocation

Sophomores: If you wonder whether it is possible to stop smoking, try this two-minute test. Close your mouth and squeeze your nostrils together until they are almost sealed. Keep on breathing. When you begin to feel desperate, squeeze a little harder-but keep on breathing?

Now let me tell you about Jenny. I saw her suffocating. Years earlier she had argued that it wasn't worth the trouble to stop smoking. Then, during a routine physical examination, they made X-ray pictures of her chest, and her doctor told her, "Jennifer, your lungs are rotten. Stop smoking!" She did. But she still had to put up with the cancer. An operation revealed that it was too close to vital lymph tissue; removal would have spread malignant cells rapidly through her body.

You should know more about that futile operation. The incision reached from just below her right breast, around her body, and up toward her back bone. Fortunately, they didn't have to remove any rib bone. By cutting through the intercostal muscles, which lace the ribs together and help in breathing, the surgeons were able to spread the ribs apart far enough to reach the cancer.

After this surgery failed, there were sickening treatments with radioactive cobalt. Then chemotherapy. When her hair fell out, she wore a wig. When the damaged lung collapsed, her oxygen supply was further depleted, and even with the help of a long tube attached to an oxygen tank, she was unable to walk across the room. When the defensive features of her blood failed, she got transfusions.

Jenny's body craved oxygen even more than it had craved nicotine a year earlier. She was eventually confined to her bed, with those plastic tubes in her nostrils. When fluid accumulated around her remaining lung, they inserted a long hollow needle and drained the fluid out. This occurred several times, each requiring a trip to the hospital.

Just as leg muscles require oxygen for action, so does the heart if it is to pump blood. The heart is mostly muscle. So is the diaphragm-the muscle that was struggling to suck in enough air when you were squeezing your nostrils.

When those muscles stopped working for Jenny, her pain stopped, too.

Kids, if any of your friends smoke and don't believe they can quit, tell them about Jenny's incision. Tell them about suffocation. Tell them to try the two-minute test.

THE HEAVY SENSE

It should be a matter of record that 88% of my sophomores, hefting two rocks, were able to decide that the one weighing 521 grams was heavier than the one weighing 473 grams.

Be it known also that only 14% of the students decided that 917 grams of sand were heavier than 857 grams of mercury.

We were concerned with the muscle sense, proprioception, and I thought we might learn more by using that sense than by talking about it. I arranged ten pairs of objects, the members of each pair differing in mass. Some of the pairs looked alike. For example, two fruit jars full of sand, one containing an empty plastic bottle that was hidden by the sand. Students moved around the room, lifting the objects and recording which of each pair was heavier.

Next day we arranged an oversize balance, visible to the entire class. It was a three-foot shaft, suspended at midpoint from a light fixture. With a noose on each end, it teetered in balance without a load. When we secured the paired objects to opposite ends of the shaft, the heavy end dipped. Although I had previously weighed each item on a good balance, the students helped with a public "weighing" and marked each others' papers. We found that for small objects (ranging from 179 to 1350 grams), there was a lot of guessing when the difference was less than 4%.

Another result was that when the objects were of contrasting density, the students often judged the denser object to be heavier, even though it was lighter. This was evident in the 14% accuracy when the sand and mercury were paired. The sand was actually 6.7% heavier. (Yes, Jim and Charlie, I assume we were determining mass, not weight. The sand was massier than the mercury). In another pair, 1350 grams of sand and 1280 grams of mercury were paired, a difference of 5.3%. Only 3% of the students made the right decision.

So now we know, we have investigated-"hands on"-what one of the sophomores called the heavy sense.

THOSE LITTLE GREEN PEOPLE

My purpose was to find out how well students would apply their knowledge of plant structure and physiology. I posed them this question:

"Readers of Comic Book are sometimes confronted by little green people from another planet. Having chlorophyll in their skin, they are able to use the sun's energy to produce their own food. Now here is your assignment: assume that such photosynthetic people really exist. Describe their appearance."

Reading their answers, I sorted out a few that deserved special recognition and read excerpts to my classes. As a final item, I read from a paper on which I had glued an article from Science Digest, December 1956, "Those Impossible Little Green Men." It was written by V.A. Eulach, reprinted from Astounding Science Fiction. It was an excellent account. It mentioned the similarity of chlorophyll to hemoglobin, admitting that mutations might adapt the blood molecule for trapping sunlight.

Much of the article, however, was concerned with the huge surface area that the green people would have to expose to light. They would need something resembling leaves. Such a large surface area would virtually preclude motion. Muscles would be of no value. To get the water and minerals necessary, the green people would depend on the soil, something like roots would be required, and so on. As Eulach said, ". . . we would no longer have an animal, let alone a man. The result would be what anyone would call a "plant," and something not too different from plants as they exist on our earth."

Reading these excerpts, I could see that students were staring at each other, wondering who was responsible. When I finished, I paused, struggling to keep my face in neutral. It seemed to me that I was improving with successive classes. But even in the fifth attempt I had difficulty. Rita broke the silence, "Okay, who wrote it?" she asked. "We hope it wasn't a sophomore."

Genes

While teaching at KHS, I managed to avoid confronting my students with the complexities of genetics. In fact, I avoided thinking beyond Mendel's peas. Perhaps because my exposure to dihybrid crosses and linkages was limited to small doses in botany and zoology courses, I failed to see much relevance for high school sophomores. I have changed my mind.

The shift probably began when I started emphasizing how little is known about human heredity. Even for traits that we know to be inherited, we generally cannot predict whether an infant will receive particular genes. And we know so little about the factors of the environment that affect the expression of those genes.

Oddly, however, parents often believe they know quite a lot about heredity. They know the potential of their children. Such parents may try to force their kids into improbable molds. This is a waste.

So, in dealing with mutations and crossovers and multiple alleles, I am not trying to train little geneticists. I only hope that our dabbling makes the students aware of the dangers of risky predictions, and builds respect for individuality.

Risk of Knowing

He was more restless than usual today, and I asked him to stay for a moment after class.

"What happened to your interest in biology?"

"I never had any."

"You must have been pretending you had."

"That didn't last very long."

"Just what's wrong?"

"Right now I don't like studying evolution. I believe in Creation."

"So do I. And I believe biologists are learning how creation takes place."

"I don't agree."

"You don't have to agree. But I would think you'd like to know more about what you are disagreeing with."

"When you believe in hell, you don't like to take the risk."

Salt, Blood, and Sophomores

Dale's face had a glow of accomplishment that I had not seen in the entire first semester. He was right on all four of the "unknowns." Solution A was isotonic, B hypotonic, C hypertonic, and D hypotonic. During the first six weeks, he struggled. The results were dismal, and he gradually relaxed his efforts. I suggested a project, trapping field mice along the railway to determine the kinds and numbers. He said, "I don't think I'd be good at that." I tried to involve him in class discussions, having told him that you don't have to be right in what you say, that there is no such thing as a dumb question or a dumb answer. But I was wrong. There are dumb ones, and because of the reactions of classmates, Dale seemed to know it.

He should not have elected biology. He didn't know what he wanted to "be," and in truth there weren't many choices open to a boy like Dale. So there he was, occupying a front corner seat, near the door. I've often noticed that in certain types of lab work an inferior student may move ahead of some who are far superior in most phases of biology. It is like that, for example, in peeling the epidermis from a geranium leaf, or in getting a chick embryo out of the shell without messing up the blood vessels.

Dale's success came in the observation of red blood corpuscles in saline solutions. The problem was to determine the relative concentration of the solutions by their effect on the corpuscles. A solution with 99.1% water and 0.9% salts leaves the corpuscles looking normal in size and shape. It is isotonic with human blood..

Water molecules diffuse in and out of the corpuscles at equal rates. If the solution contains a higher percentage of water and a lower percentage of salts, the solution is hypotonic with blood. Water diffuses into the corpuscles faster than it diffuses out. The corpuscles swell. If the osmotic pressure is great enough, they burst. The red pigment leaks out, leaving pale "ghosts" instead of corpuscles. And if the concentration of water is lower than 99.1% (as a result of more than 0.9% salts), the corpuscles lose water and shrink. Their membranes become rough, with peculiar little bumps.

I doubt that Dale understood what he was seeing. Certainly, he did not see the broad significance of osmosis. It is part of the process of keeping organisms in balance with their environment. It affects the intake of water by roots. It explains why a freshwater fish must keep expelling excess water to prevent dilution of its body fluids. It explains why a marine fish must keep drinking and excreting salt to prevent dehydration. Medical laboratories sometimes run a "fragility test" on a patient's red corpuscles, to determine how hypotonic a solution must be to break the membranes. This is a clue in diagnosing certain disorders.

When I asked Dale why the corpuscles in his solution A were larger than those in C, he was stopped. Looking at his notes, he replied, "I guess it's because my blood is isotonic." A bit twisted, but he used a good word.

One of the features of this lab work that I like is that each student works on his or her personal unknowns. There is no leaning on a lab partner. Another feature is that I can score students in an unusual way. If they make an error, they can try again, using a new set of solutions. The final score depends on how many trials are necessary. Very few students want to quit before getting all four of the unknowns right.

When I first tried this lab problem six years ago, the results were so poor that I was tempted to scuttle the whole idea. Troubles began when I attempted to make a solution that was precisely isotonic with human blood. I jabbed my fingers in so many places I was about to run out of fingers. The shortage of microscopes, one scope for two students, resulted in a time problem. Some students had trouble reaching blood, perhaps their lancets were too short. Others fouled the high-power objective lens with bloody saline and reported that the microscope wouldn't focus.

I had trouble convincing students that too much light would virtually erase the corpuscles from view. This was particularly serious in a strongly hypotonic solution; those pale ghosts just disappeared. This led students to believe there wasn't enough blood on the slide. More blood wasted, and more time.

Another difficulty arose when students spent too much time staring at their preparation. Water evaporated from the edges of the cover slip, increasing the salt concentration. A solution that was isotonic when placed on the slide became hypertonic. I've forgotten who it was, but one girl allowed so much water to evaporate that salt began to crystallize. She thought her blood was "real crazy."

All told, the tests were disappointing, and I didn't schedule that work the following year. By the time the next set of classes came around, we had a microscope for each student. I reviewed my notes and made more solutions. Thinking that a medical technologist might have some useful advice, I called Al Harris.

Al's first recommendation: "Use expired blood-bank blood. Then you'll have no doubt how the cells are going to respond." To back up his suggestion, he came to my lab with a pint of expired blood. "It's just beyond the safe shelf life," he said. "It should be good for an additional week or more if you keep it in a refrigerator."

Another recommendation from Mr. Harris: "Keep the ratio of blood to saline low."

We tried various ways of standardizing the procedure and finally I came up with an easy one to follow. One small drop of saline from the dropper bottle, and one toothpick tip of blood. We found that if we dipped the small end of a toothpick into the blood about a quarter of an inch and allowed the excess to drip off, there were plenty of cells left, and they were easily mixed into the saline with the toothpick.

After Al left, I checked one of each of the unknowns. The corpuscles behaved properly. Since all of the isotonic samples were poured from the same bottle into clean dry dropper bottles, there should be no problem. The same for each of the other two solutions, both of which are less critical than the isotonic.

Next day the students had a little time for a final look at the labeled "knowns" if they wished. When satisfied, they selected their personal numbered set of test solutions. For example, in set number 1, the dropper bottles were labeled 1a, 1b, 1c, and 1d. The students knew that each set would contain one of each of the three solutions plus a second sample of one of them.

There were enough problems with microscope lights and fouled lenses to keep me busy, but everyone completed the tests on time. The results were far better than the first year.

In the years that followed, we ran into trouble with some of the expired blood. Too many corpuscles were misshapen. I don't know why. That is why I began recommending that the students check their samples with their own blood as well as the blood bank supply.

When salt, blood, and sophomores got together today, instructions were refined by frustrations of previous years. Dale picked up sample set number 11. He examined each of his solutions with the expired blood, then checked one of them again with his own supply.

Although I am able to check most of the results as they are turned in, I wait till the next day to reveal the results. I use the same sets of unknowns for all of my biology sections, and word might get around about the answers. For Dale, however, I couldn't wait. I motioned for him to stay as the remainder of the class left. When I told him that his decisions were all correct, he let out an embarrassed laugh. Then he said, "You know, I thought maybe I'd do it right this time. That's good!"

It was good for me, too.

The Two Ends of a Bee

I have just made one of those lab-shaking discoveries that overhaul the teaching of science now and then. The problem: how to persuade sophomores to look at the leg of a bee and draw it as it is, instead of copying the hasty sketch that I have made on the chalkboard.

The solution: stop making those hasty sketches on the chalkboard.

Well, it's really not so simple. The antenna cleaner is elegantly designed. Students should see it. They should figure out how it functions.

When the antennae become fouled with dust or pollen, the bee is deprived of one of its major sense organs. It responds by reaching forward with a front leg, which is equipped with a remarkable tool. At the upper end of the tarsus, there is a cup-like notch, lined with a brush of minute spines. Above the notch, at the lower end of the tibia, a claw-like appendage extends downward. When the bee fits the base of its antenna into the notch, the tarsus is flexed, moving the "claw" downward and locking the antenna in the notch. By extending its leg forward and downward, the bee scrapes off the offending dust. It is an elegant device.

It is too small to see without magnification, however, and a clear view requires precise alignment and lighting. Failing to anticipate the glow they would feel in coming close to a real antenna cleaner, students are inclined to copy what they see on the chalkboard. I found, however, that I could not dispense entirely with the chalk. Some students have trouble finding the front leg.

My discovery was this. I sketched (hastily) the outline shape of each of the bee's three pairs of legs, marking the location of various structures: antenna cleaner , wax probe, pollen brush , pollen press, and pollen basket. The location, not the design.

Knowing where to look, but having no drawing to copy, students observed more carefully. Some of them needed to be assured that they were looking in the right direction. Denise, for example, was having trouble finding the barbs on the worker bee's sting. She had removed the structures to be studied and had lined them up on a sheet of paper. A good start.

"I can't find those things that keep the sting from pulling out easy," she complained. "Do you think mine maybe was a drone?" "I think maybe the drone is a male," I said. "Where is the sting on a male?"

She remembered. "But I can't find those little things!" she repeated.

"Denise," I explained, "that structure you have on the microscope looks more like a tongue than a sting. Are you sure..."

"Well," she decided, "if it's not a drone, maybe I have the wrong end of the bee."

THE SHRUG

One of the distressing aspects of teaching is to find a capable student who is willing to steal a grade. When I scored a set of tests, yesterday, Edward's was missing. Today, while returning the papers, I asked why he hadn't turned one in. He said he had. While I continued distributing the tests, Edward went to the front of the lab and looked among the things that were cluttering the worktable. In a moment, he came back to me and handed me a paper.

"I guess you mislaid it," he said. "It was under some magazines."

I thanked him and said he could stop after class and see his score.

After getting the lab work on stem structure under way, I checked his answers. They were all correct, far out of line with Edward's usual performance. Besides, the paper quality didn't match the blank sheets I had distributed for the test. I marked the paper with a perfect score and left it on the table.

At the end of the period, Edward came by, inspected the paper and said cheerily, "Hey, this is better! You know I was about to get grounded if I didn't bring my grades up."

"Yes," I said, "but you know, Ed, I can't record that score. Where is the answer sheet I gave you at the time of the test?"

"I don't know what you mean," he replied. "This is it."

"No," I returned. "We never distribute bond paper for tests. You must have written the questions on the answer sheet and taken it home. This paper was not on the table until you put it there sometime today."

Edward hesitated, then shrugged and walked away.

Shadow Problem

Do shadows indicate danger for mosquitoes? I'm still not sure. There is no doubt, however, that we learned something trying to find out. And it wasn't intentional, at least when it started.

Come to think of it, many student "discoveries" aren't intentional. They just happen, because kids are looking at something alive. No purpose. No proper rationale.

That's how it was with the wigglers. Randy M. found them in a plastic bucket of rain water. He brought them in to feed the Squaw Creek minnows that populated one of our aquaria. For temporary storage, we poured them into a battery jar. And to invite observations, we placed a low wattage lamp above the jar.

Students were back there watching, noticing that although the wigglers were generally at the surface, occasionally one of them wiggled its way to the bottom. Someone was arguing that mosquito larvae are air-breathers, and that's why they don't remain submerged. Then for some unknown reason Randy reached over the shoulders of other watchers and switched off the lamp. The response of the wigglers brought an excited gasp from the watchers. Both larvae and pupae immediately submerged, wiggling their way to the bottom of the jar.

"They're afraid of the dark!" said Randy, motioning for me to come and watch. Randy has quick opinions. And though he seldom resolves an issue, he often keeps a discussion moving with his little fillers while more cautious students assemble their thoughts.

I believe it was Randy who earlier in the fall suggested why the ailanthus is known by such contradictory names as "tree of heaven" and "stink tree." He said it's because the tree stinks to high heaven.

Liz countered Randy's fear-of-darkness notion, suggesting that the wigglers may be attracted to light, the way moths are, when the light was turned off, there was no reason to remain at the surface.

Although I had still another hypothesis, there was only a minute of class time left, and I remained neutral. That gives me time to look for evidence in print, and that's another plus for unplanned observations. It's more exciting to search for an answer to your own question than merely to inform yourself in general.

After school, I moved the battery jar and light to the demonstration table, where it would be visible to the entire class. After the wigglers had settled down (settled up?) I tried something I'd been thinking about while monitoring the lab work on grasshoppers. With a continuous motion, I passed my hand between the light bulb and the jar. The momentary shadow sent most of the wigglers into a dive.

Protection from enemies? If so, what enemies?

After recording a set of test scores and filling out several mid-term warning slips for the parents of reluctant learners, I pedaled home, thinking about mosquitoes. We need a common name for the pupae. They don't really wiggle. They're shaped like commas, but if we called them commas, we'd have to find a grammatical name for the larvae. Exclamation points wouldn't do! If pupae don't wiggle, how do they move? Well, they tumble. Somewhere, I suppose, I've heard them called tumblers.

OK: egg, wiggler, tumbler, bloodsucker.

No, bloodsucker isn't quite right. The female sucks blood, when she can get it. In some species, a meal of blood is required for egg production. But the male is a nectar sipper. He probably couldn't penetrate my skin if he were inclined toward blood. I once caged three males and offered them an over-ripe Hawthorne fruit, but they ignored it. When I punctured the tiny apple with a needle, making the cider easier to reach, one of the mosquitoes sucked until its belly bulged like a balloon.

The shadow - that was mainly the thought that prompted me to check a few authors that I found on my home shelves. Marston Bates might have a clue, but he wasn't represented. I was soon lost, however, in L.O. Howard's Mosquitoes, published in 1901. When the chapter on mosquito enemies failed to light up the shadow problem, I lost myself in the fascinating accounts of research on malaria and yellow fever. At the time Howard was writing, the implication of mosquitoes in these diseases was exciting news.

One of the events widely reported in the autumn of 1900 concerned two doctors, Sambon and Low, of the London School of Tropical Medicine. To test the bearing of mosquitoes on malaria, they built a tightly-screened house in southern Italy, where malaria was rampant at the time. They lived there during the most hazardous season, taking no health precautions other than remaining in their mosquito-proof house from dusk to daylight. They exposed themselves to drenching rains, which many people believed would surely bring on malaria. Their windows were open, exposing the doctors to the "bad air" for which malaria was named. In the evening, they waited by the screens, recording the arrival time of the anopheline mosquitoes.

While many unprotected people were burning and chilling, Sambon and Low remained healthy.

Now this did not prove that mosquitoes dispense malaria. When nothing happens, you don't know that it couldn't happen!

Nevertheless, the experiment was important circumstantial evidence. And back in London a young man named Manson volunteered for a more positive check on the Sambon-Low test. Far away from the malaria country, he allowed a mosquito that had fed on a malaria patient in Rome to help herself to his blood. After the expected incubation period, Manson developed the dreaded symptoms. And when his blood was examined, it contained the protozoa that were always associated with tertian malaria, the organism that we now know are delivered by anopheline mosquitoes.

Dr. Howard's book reports also the heroic experiment on yellow fever by Dr. Walter Reed's team of physicians. One member, Dr. Jesse Lazar, died during the tests of yellow fever. It is a familiar story, but because I had been thinking about mosquitoes most of the day, it had special impact.

I could find nothing about the shadow dive, however, so I checked the university library. Fortunately, The Natural History of Mosquitoes (1949) by Bates was there, and I found answers for many questions that I hadn't asked. For example, "The pupa in most mosquito species is considerably lighter than water, its buoyancy due to a large air space between the wing cases on the underside of the cephalothorax. By vigorous movements of the abdomen, which terminates in two large paddles, the pupa can propel itself with considerable speed, though without control of direction, gaining depth by a series of somersaults." Another gleaning: you can watch dragonfly nymphs gobbling up immature mosquitoes in a shallow aquarium, and you assume that this happens also in ponds and streams. But Bates says that J.B. Smith took care of this misconception in 1904. Smith had found that dragonfly nymphs are bottom feeders, where they have little or no opportunity to prey on mosquito larvae.

When I located Bates' section on alarm reactions, I was getting closer. I learned that some kinds of mosquito larvae will dive when the reduction of light intensity is so slight that it is "hardly detectable by the human eye." In one species, Culex pipiens, the wigglers are so strongly attracted to light that if a glass container is lighted only from the bottom, the wigglers in it will remain submerged and drown. (Are you listening, Liz?)

But that doesn't explain why Randy's wigglers were diving. There was no light down there to attract them. My own assumption was that the natural enemies of the wigglers and tumblers have more difficulty seeing their prey against the dark bottom than against the brighter surface. Because the immature mosquitoes are air breathers, they must generally remain at the surface. The shadow dive, I thought, protected the species from surface-feeding fish and predaceous insects.

Next day in class when I asked if there were other ideas about the dive, the students were blank. Those whose heads were pointed in my direction looked with glassy eyes.

I said, "I am your biology teacher. This is Wednesday. Yesterday was Tuesday, and some of you saw the mosquito wigglers dive to the bottom of the jar when Randy turned the light off. See, here are the wigglers, and this is the light!"

Randy nodded feebly, and Liz finally opened by repeating her light- attraction idea. So I knew that I was in the right classroom.

It was a slow start, but after I had flipped the switch and the wigglers obediently went to the bottom, ideas began to flow again. Marjorie P. proposed the protective adaptation, much as I had considered it. And Scott O. soon repeated it, as though it was his own idea.

The class generally was ready to settle for the protection hypothesis. There was one aspect, however, that troubled me. "Where would a predator be located in order to cast a shadow on the larvae?" I asked.

Scott said he agreed.

"He didn't say anything to agree with!" Marjorie retorted. "He asked a question."

"I agree with the question," Scott snapped. "It would have to be a bird or something."

There were still too many blank faces in the room. It's that way, at times. You assume that kids are thinking about the same thing you are.

"Well, before we go back to grasshoppers," I said, "listen to one sentence that Marston Bates wrote. He's an ecologist at the University of Michigan, and he has done a lot of research on mosquitoes. This is his book on mosquitoes, and here is one good sentence - listen: 'Thus in studying mosquitoes and predators we come, as always, against the problem of our ignorance "

It is comforting to find that someone like Marston Bates didn't have an answer.

On Digging Ditches

If I want my students to keep respectable notebooks, I must be willing to spend countless hours reading and marking. And I must be willing to follow this with hours of personal counseling. I am not willing; I am frustrated.

In my classes, a "satisfactory" notebook is required for a passing grade. I explain that no one should be required to memorize all of the useful facts and ideas of biology; some should be stored in a notebook. And I try to demonstrate that systematic note-taking is a way of learning.

Before reporting grades, I sort notebooks into two piles, satisfactory and un. Because much of what I see is painful, I consider the merits of digging ditches.

I could enforce a superficial improvement by awarding points - five for neatness, another five for dating all entries, two for properly spelling nucleus, more for telling how you conducted the experiment, rather than copying the instructions from a lab manual - sophomores love points.

To back up my assertion that some kinds of information should be stored in class notes, I occasionally allow use of notebooks during a test. Anything on the notebook pages, in the student's own writing, is acceptable.

When students bring their papers to my desk, I sometimes ask, "Did your notebook help?"

A good student may respond, "I didn't need it."

I reply, "Maybe that's because you had a good notebook. You organized your thoughts in advance."

A poor student is likely to bristle and say, "You asked the wrong questions!"

Advocates of meticulous scoring may say that I haven't given this approach a fair test. They may be right. But if a decent notebook isn't its own reward, I have already failed. I should dig a ditch.

THE DANGER OF BEING RIGHT

In my first years of teaching, I must have been dimly aware that authoritarian tactics are opposed to science. Surely, Starrak and Hamlin and Sass got through to me in some degree. But looking through a sequence of tests that I wrote, I suspect that I gradually rediscovered the danger of being right. Students who are dominated by authoritarian teachers learn the verbiage of science, but they are indoctrinated with anti-science.

As this became clearer to me, I used less class time for "recitation", which is generally dull, and more time for inquiring types of discussion. My aim was for students to see that most problems have more than one possible solution. None of the solutions may be fully satisfying, and the method of investigating a question may be more important than the answer.

This looks great. But there is a barrier to success. It is so much easier to test for little, isolated facts than to find how well students can use those facts.

With uncomfortable insight, Nels A. put it this way: "You tell us to think. When you give a test, you find out if we know what's in the book."

Ouch!

WITHOUT A HORSE OR A GOOSE

A great idea today: For introducing the study of blood pressure, we'll lead an old horse into the lab, expose its carotid artery and connect it to a long glass tube. If there is any problem making the connection, we can use the trachea of a goose as a link. Holding the tube straight up, we'll measure the height to which the blood rises. This will represent the pressure of the contracting left ventricle.

I was about to start looking for a horse and a goose when I realized the flaw. Stephen Hales did this in 1733, and the blood ascended 9 1/2 feet. The ceiling of our lab is too close to the floor...back to the sphygmomanometer.

But the idea that came while I was watering the plants in the greenhouse didn't require such a high ceiling. I was using a hose with a mist nozzle. As had often happened, the spray continued to function for a while after I had closed the faucet.

That was it! For a demonstration of diastolic pressure I could use a long rubber tube at the laboratory sink. A nozzle made from a short length of glass tubing, heated and drawn out to a small opening, would restrict the flow of water, just as I had observed in the greenhouse. It worked.

Without announcing the nature of the analogy, I asked the students to help explain a puzzling situation. Turning the faucet lever slightly, I held the glass nozzle firmly and directed the jet of water into the sink.

"Now," I said, "I'm turning off the faucet. There, it's off." The water continued to squirt. It squirted for half a minute. When I asked for an explanation, one student thought it was some sort of Trump trickery. So, I asked him to come up and repeat the operation. He did, and the water did.

Most of the explanations were vague. "The water has momentum, it can't stop moving right now!" "I think it's air pressure." "It's something like a siphon. It's like gravity is pulling more water out."

In every class, however, someone suggested that when the faucet was open, the pressure of the water stretched the rubber tube. When the faucet was closed, the expanded tube squeezed water out until the tube's diameter was back to normal. For verification, I asked a student to hold the tube firmly with one hand, the glass nozzle with the other, while I opened and closed the faucet. Right. The tube expanded, then contracted. He could feel the change.

Now, I explained, you know something about blood pressure. Your arteries are like the rubber tube. They are stretched by the pressure resulting from your contracting ventricles, by your systolic pressure. When your ventricles relax, the muscular walls of the arteries contract, squeezing the blood forward. That's your diastolic pressure. All without a horse or a goose.

Streptomycin Treatment

"We love you Trumpy-poo!" For several days, we've been observing the inhibition of three bacterial species by various antibiotics. With the measurements pooled from each of my five classes, I had summarized the data on the chalkboard. While the students worked on their reports, I disappeared in the direction of the beehives.

When I returned, the kids abruptly returned to their reports, trying to look as though they'd been working all the time. Fortunately, I noticed the love note on the chalkboard as I came in the door. Pretending that I hadn't seen the note, I asked if there were any questions. Did they remember that in summarizing the data they could round off measurements to the nearest whole millimeter? There was the usual pause. Then, trying to hide a grin, David E. asked, "Mr. Trump, are you sure those figures on streptomycin are right?"

Without turning to look, I said, "I know you love me, Dave, and I really appreciate your saying so."

Dave's expression slipped into reverse, and the class broke into laughter. "I didn't do it, Mr. Trump. I really didn't!" "You shouldn't be embarrassed, Dave. I think it was nice of you to express the sentiments of the class. Are there any other questions?" I waited for the class to leave before erasing the greeting.

SPEEDWAY SCIENCE

Bruce came home with his quiz paper from Lindy Solon's junior high science class. On the previous day, the students had been tested on a film concerning the Indianapolis 500. Of the 31 questions, Bruce had answered less than half. Here are samples questions:

How many cars were entered on the first day?

How many people attended the trials?

What product was advertised in this movie?

What was the age of the oldest driver?

In what city were the races run? (Bruce was right on this one.) I predict that Solon will move on to a higher-paying job.

IN A NUT SHELL

Every biology laboratory should have a copy of Experiments with Plants by W.J.V. Osterhout, published by Macmillan. Here is an abbreviated example: Place seeds such as English walnut in a solution of eosin dye. Remove them at intervals and examine to see the path of water penetration. In seeds such as walnut and pecan the hardness of the shell is offset by "an absorbent, wick-like, central strand which takes up water directly from the opening." Unfortunately, the book will be difficult to find. It was published in 1906.

I'VE BEEN OVER-TEACHING!

Despite the notion that learning to identify a bird or a tree is not really science, I continue to help students know who's who in the natural world. I suspect that proper identification is at least one of the tools of science.

With the expectation of becoming a forester, I learned to recognize most of the native trees while still a high school student. My guide was a book by Harriet Keeler. Later at Iowa State, classes in systematic botany and forestry extended my grasp. Then, back in Keokuk as a novice biology teacher, I led my classes to some of the same trees that I had known as a student. But with more experience,

I believe I was over-teaching, and I continued to do so for a few years after coming to Ames High. It was a showing and telling approach, not very efficient. We moved from tree to tree, waiting for the stragglers to catch up. Sometimes we got back in time for the next class.

I am uncertain just when I began placing more responsibility on the students, but now I do not "conduct" this fieldwork. We begin with an indoor lab session, examining typical leaf types and arrangements. Trees such as poplars, elms, and all but one of the maples have simple leaves- one leaf blade fastened to the twig by a petiole. A bud, which will produce next year's leaf, marks the node where the petiole begins. Hickory, ash, and boxelder (a maple) are examples that have compound leaves; they have multiple leaflets on one petiole.

Some trees, such as ash and maple, produce leaves in pairs, opposite each other on the twig; and one Iowa native, the catalpa, arranges its leaves in triplets. Others such as elm, poplar, and hickory produce only one leaf at each node. They are staggered on the twig, alternately arranged. I do not identify the specimens that the students are examining. But with the help of sketches on the chalkboard, they can decide that this one has simple, alternate leaves, another compound and opposite - and so on.

Armed with their sketches and notes, the students fan out into the assigned area with a duplicated map. It tells them that, on the corner of 7th and Clark, there is a labeled silver maple; at the east end of the Squaw Creek bridge, a cottonwood; at the west end, a Kentucky coffee-bean tree. With the expectation of an identification test or two, the students make the rounds, taking notes as they see fit. They know that I will be commuting through the area, offering help and seeing that the boys don't throw Nancy off the bridge.

How does it work? It is easier for me and more productive for most of the students. They appear to enjoy it more. And because they know it is their responsibility, not mine, they generally return on time.

Goose Raising, Crawl Spaces, and How to Fight It

Each month the Government Printing Office sends a listing of new publications. Yesterday, baffled by the obstacles to reasonable teaching, I considered applying for a job in Washington, writing reviews for current bulletins. To make sure I could handle such an assignment, I practiced on a few recent listings:

ABSOLUTE VISCOSITY OF WATER AT 20 DEGREES C. If you are concerned about sticky water, this is absolutely your best buy.

HORN WORMS ON TOBACCO, HOW TO CONTROL THEM. If the recommended insecticides fail, try nicotine.

THE WHEAT JOINTWORM, HOW TO FIGHT IT. This eight-pager explains that the jointworm is not a worm. To find what it is, Order A-1.35: 380 for 5 cents.

WOOD TICKS, HOW TO CONTROL THEM IN INFESTED PLACES. Wood ticks are particularly abundant in infested places. Those are the best places to control wood ticks.

BEANS, BULLETS, AND BLACK OIL. A bulky book printed for the U.S. Navy, with a title that may have been coded to confuse the U.S. Army.

SHELLFISH WITH CERTIFICATES, CRAWL SPACES, THE STEEL SEIZURE CASE, GOOSE RAISING, and CULLODEN IMPROVES ITS CURRICULUM are all the best in their respective categories. So are SANDWICH CONSTRUCTION FOR AIRCRAFT and INFLUENCE OF YARN SIZE ON THE RELATIVE CONTRIBUTIONS OF SIX COTTON FIBER PROPERTIES TO STRENGTH OF CARDED YARN.

HISTORY OF PLAGUE IN THE UNITED STATES. What most of us call epidemics are really endemics that smolder through a low phase, then flare into pandemics.

SARATOGA SPITTLEBUG. If you think the Saratoga Spittlebug is a popular new dance step, you are wrong. The Saratoga Spittlebug is an unpopular old insect that spittles around in the vicinity of Saratoga.

If my offers to Washington are not appreciated, there is another option. The world is waiting for a popular new dance step. By observing one of the evening celebrations in the Ames High study hall, I could find something that needs only a dignified name. How about Spittlebug Bash?

GEYSER

Every teacher needs a day like this one. I take five or more raw eggs to school. If I am reasonably coordinated, we can cover an important lesson without mentioning what it is. Then tomorrow the lesson will gradually emerge as the students try to explain what they have observed. "This is an egg," I say, holding one up for view, "a chicken egg, a raw chicken egg. How can we remove the shell?"

"Boil it and peel it."

"We want a raw egg."

"Make a hole in each end and blow the insides out." "We want a whole egg, one without a shell, that looks like an egg."

Eventually someone remembers the directions that came with a chemistry set: "Soak it in vinegar. The shell will dissolve."

Okay - acetic acid. In fact, any acid. We settle on hydrochloric because I happen to have a bottle of dilute HCl right there in the sink.

"But before we remove the shell we must know the volume of the egg. Any suggestions?"

"Pour it in a beaker."

"You've forgotten. We want it whole, like an egg."

There are blank looks at first. Then one of the wise ones says, "Measure the height and multiply by the hypotenuse or something." That gets a laugh along with the admission that geometry, even the solid variety, isn't going to help. But finally, someone remembers that in junior high science they determined the volume of irregularly shaped objects by finding how much water they displaced. Good! I happen to have an overflow can in the sink. Placing it at the edge of the sink, I pour water in and let the excess flow out of the spigot. Then I lower the egg into the can and catch the overflow in a graduated cylinder. When the last drop has dripped, I hand the cylinder to the nearest students. The two of them take turns squinting at the level of the liquid, disagree, and bump heads trying to look at the same time. Finally, they nod to each other and announce that the volume of the egg is 59 ml. I write this on the chalkboard, along with the date and time.

"Now what is your estimate of the volume after the shell has dissolved?" There are several guesses ranging from 52 to 58 ml. If anyone suggests a volume greater than 59 ml, I try to look stern and ask, "Are you having a bad day?"

Then, without comment, I pour a little acid into a beaker of water and cautiously lower the egg into it with a loop of wire that happens to be there. And while the students are watching bubbles rise from the surface of the egg, I write an equation on the chalkboard:

$CaCO3 + HCl \sim CaCl2 + H2O + CO2$

If there is enough time, I fail to indicate that there must be two molecules of hydrochloric acid for each molecule of calcium carbonate, and I ask the class whether it is a balanced equation. If not, how do we balance it? We indicate two molecules of the acid.

Next, I describe a bit of homework: Use two carrots of about the same size. Cut off the ends so you have pieces three to four inches long. Carve out some of the central tissue, making two little waterproof cups. Carve as deep as possible without cutting through the bottom; they must be waterproof. Pour sugar solution (or pancake syrup) into one carrot, half full. Set it in a bowl of water, with the water level even with the level of the syrup. Do the same with the other carrot, except that you reverse the two liquids-water in the carrot, syrup in the bowl. Place the bowls where they won't be disturbed. Inspect them before retiring, again in the morning.

As the period is about to end, I invite anyone who has free time to come and measure the egg. Several respond, and when they discover that the volume has increased, I suggest that they keep it secret; let others discover for themselves.

Next day, before bringing out the egg, we hear reports on the carrots. By morning, we are told, the carrot containing syrup was overflowing. The other was almost dry. Without considering an explanation, we check the volume of the egg. Finding that it has almost doubled, we look for similarities between the two experiments. Someone mentions osmosis. But that is only a name. What's happening? Why do water molecules move through carrot tissue towards a sugar solution? Why do they penetrate an egg membrane into a mixture of complex organic molecules?

At this stage, I settle for understanding that water molecules diffuse faster from a place of high concentration (of water) than from a place of lower concentration. When membranes separate the two fluids, a pressure develops. This is osmosis. Active transport will wait.

In the meantime, it is good to know that fresh water fishes must excrete water to keep from blowing up; it comes in through their gill membranes. Marine fishes must drink water, and excrete salt, to keep their tissues from drying out. It is good also to know that the egg membrane is relatively tough. I drop the egg on the tabletop, from a few inches. It bounces. Ignoring suggestions that I drop it from a higher level, I place it in a bowl.

After assigning a notebook report on the observations, including drawing a graph to depict the time-volume relationship, there is one last question: "What will happen if I puncture the membrane of the egg with a fine insect pin?" With a pin in hand, I don't wait for an answer. A tiny geyser erupts. The thin stream of water goes 18 inches high, perhaps 19. The stream continues for more than a half minute, gradually subsiding as a geyser should. And I have a good day.

Skulls, Cocoons, and Diabetes

I found the skull on an eroded bank of the Skunk River. Wanting to verify that it had belonged to a bison calf, not a domestic cow, I showed it to Tom B. Tom is a sophomore. He finds bones and learns where they came from.

When I wanted to know when to expect the emergence of a cecropia after I brought the cocoon indoors in January, I asked Paul P. He, too, was a sophomore at the time of my question. He studies the cecropia.

And when I wanted to know more about the symptoms of insulin shock, I stopped Maureen M. after class. She listens when her doctor explains the physiology of diabetes.

I wish I had kept a list of sophomores who know more about some biological topic than their teacher. There have been others. This thought hit me when I saw a sign on a farm gate: "I repare hydrolics." Mentally I marked the repairman down on spelling. Then it occurred to me that if my hydraulics were faltering, he is the one I would call for help.

TRY AGAIN?

I was in the supermarket line when Alice told me her married name and asked if I remembered the time when she tried to throw a ball of paper into the waste-basket.

"I missed," she explained, "and it was right at the first of the period. You had already started with instructions of some kind, and I didn't think you'd seen me throw. But a few minutes later you walked over while you were talking and picked up the paper. Do you remember what you did?

I didn't, and Alice continued, "You laid it on your desk. I was afraid you would straighten the paper out and see the notes Jan and I wrote about our dates. Well, when class was almost over you brought the paper back and handed it to me. You don't remember? You handed it to me and said, 'Do you want to try again?'"

TRAINING THE TRUTH-SEARCHERS

September is a wonderful month. It has crickets and red sumac leaves. It has hot winds and frosty nights. It has grackles assembling for migration, footballers gathering for bruises. And sophomores.

I have read that science is the search for truth. If so, then science teaching must involve the training of truth searchers. I wonder. Is this what I am doing?

The first day of school, I assume, should set the tone for what is to follow. One September this idea prodded me, and I was prepared. When the students spilled into my room and vied for the rear seats, there was a living cricket in a petri dish on each table. I said nothing about the insects. After distributing the books, I explained the importance of a good notebook, saying that at times students could use their notebooks during a test. Next day the crickets were gone, and I asked my searchers to write a brief comment about what they had observed in the petri dish.

Many of them looked at me in astonished disbelief. They had not been told to look in the dish. So I advised: If you have no curiosity about living things, you are in the wrong course. See your counselor.

This was the measure of their curiosity-and their spelling: "It had two long fellers in front and kept waving them around, looking for a way out...I think it was a black bug...The back legs were bigger and were used for hoping...A black thing with three stingers on its tail...I saw a cricket living without fresh air."

Those who are able to think like a cricket wrote like this: "When I whirled the dish it felt dizzy and stopped moving...he didn't like the idea of being captive. He had six legs and wanted very much to get out and use them." Should I explain the meaning of anthropomorphic?

This year I postponed some of the clerical work for a couple of days, allowing more time for biology. On each table was a culture bowl of pond water and several whirligig beetles. The assignment: "Learn all you can about whirligig beetles." No pages to read. No charts to fill in. No outline to follow. Learn about whirligigs!

This approach, I assumed, would suggest that the proper reference material for learning about living things is living things. We spent the last ten minutes of the period hearing comments from students about their observations. Many of them concentrated on anatomy - size and color, number of legs, streamlined design. Others were concerned mainly with behavior - crazy patterns of movement, diving, climbing to the edge of the dish and flying to the windows. Some of the students had more questions than answers; that was good, too.

The disturbing observation was mine: Many of the students needed only ten minutes to learn all they could about the beetles.

Next day I distributed textbooks and allowed time for reading the first few pages, about whirligigs. Then another opportunity to learn from the living insects. This time, more of the students saw the air bubble at the tip of the abdomen, the source of oxygen during a dive. They recognized the striking difference between the front legs and the two rear pairs. And with the help of hand magnifiers and binocular scopes, many of them appreciated the advantage of the divided eye, the lower half for underwater vision and the upper half for views above the surface.

As one student put it, "Yeah, I see it now. I suppose it was there yesterday, too!"

An important part of the lesson: It is great to discover, but there are times when it is good to have a little help. Reading and listening are part of searching for truth.

How to Dewitch a Waterwitch

When the debunking of superstitions came up, Steve Knudsen's story was another example of. "I wish I'd thought of that!"

Before heading the film production unit at ISU, Steve taught science, and he felt an obligation for defusing any superstitions that prevailed among his students. In a preliminary lesson he found that, although his 9th graders were not afraid of number 13 or of breaking a mirror, they were divided on the authenticity of finding underground water with a forked stick.

"Let's investigate this," Steve suggested. They located a local water-witch; a tobacco-chewing water-witch. They followed him around an alfalfa field until finally his forked stick bent down. "This is where you'd find water," said the witch, punctuating his pronouncement by expelling a brownish glob of saliva.

As the students walked away, they secretly marked the spot; this was part of their plan. Now the test: They asked the witch if he could find the same spot while wearing a blindfold. He agreed to try, but added that there might be other locations for well-drilling in the same field. So they led him around the field in a circuitous route, keeping a safe distance from his tobacco juice.

Eventually his stick turned down again, and the students eyed each other in wonder. It was precisely the spot where he had first announced water. There was stunned silence from both Steve and his students. Then one of them said, "I think we should let him try again."

So they tracked confusing circles around the field again. And again, the stick bent at the original spot!

Now I doubt that Steve Knudsen had been taught how to handle a situation like this. It's not in the manuals. When I asked how he recovered, he replied that he didn't. But one of his students came to his rescue by saying, "But we still don't know there's water down there!"

More Demons

I suspect that ours are the best-fed ant lions for miles around. Students vie to be first in the room with fresh flies.

William Morton Wheeler called them demons of the dust. I have found most of mine in sand, in the dunes of southwest Michigan and beneath over-hanging ledges along the Mississippi River bluffs. A few of them dig their pits along the foundation of our house, where the wide overhang of the roof keeps the soil dry and dusty. There aren't enough here, though, to provide for lab observations. That's why I brought about two dozen of the Michigan specimens back to Iowa.

For years I have wanted to expand the observations that I began during my Panama time in the U.S. Navy. That's why we were housing ant lions in trays of fine sand at the back of my lab. In a 20 by 25 cm tray (the size of an 8 by 10 inch photographic print) we placed four ant lions. Next morning, after the four pits were complete, we photographed the tray of sand, sifted out the larvae, added one more specimen, and placed all five of them in the middle of the tray of sand. Each day we added one more larva and photographed the pattern of pits. By projecting the negative onto an 8 by 10 inch easel, we could easily study the spacing, life size.

Measuring the distance between the two nearest pits, we found that this figure varied from 30 millimeters to 82 millimeters-and it did not appear to be related to the number of larvae in the tray. If the lions were spreading out as far as possible, this distance would decrease as more larvae were added to the tray. They would be crowded closer together. This didn't happen. When only four larvae were in the tray, the nearest two pits were only 31 mm apart. With eight in the tray, the shortest distance between two pits was 38 mm. It is, of course, possible that there is no crowding factor until the distance is less than 30 mm. More tests.

Another problem: When the number of larvae reached eight, some of them did not dig a pit. With 12 larvae, we found only eight pits. And with 15 larvae, we found only seven. Why this drop-off? Was the daily disturbance of their traps responsible? When we have another batch of ant lions, I want to reverse the sequence, beginning the test with 15 and removing one each day. More tests!

Occasionally, I am asked why I continue teaching biology. Perhaps it is because I still don't have any of the answers.

On Leaving the Woods

"I left the woods for as good a reason as I went there. Perhaps it seemed to me that I had several lives to live, and could not spare any more time for that one..." Thoreau wrote that after leaving his cabin at Walden Pond. His statement, typed on a filing card, has been on the bulletin board above my typewriter for years. It prods me.

I stared at it last evening, after Lorene told of a conversation with a member of the Board of Education. "Why does your husband stay at Ames High?" the board member asked. Lorene told her that I like it here, mentioning a number of things that make Ames High attractive.

It is true: Squaw Creek, with its minnows and whirligig beetles is a few hundred yards from home; a pine grove with horned owls is only a quarter of a mile north; a new high school is being planned for the edge of a prairie a little farther north; and Iowa State's fine library is only ten minutes west.

Yet I have known other spots on the earth that I would surely like, too: I found the clear streams in Pennsylvania desirable; also, the loon-call lakes of Minnesota; I liked the edge of the redwoods in California, and the majesty of the mountains in Montana; even the surf-smell of the Gulf of Mexico. I could become involved with marmots in the mountains as well as with woodchucks; with sea urchins as well as with honeybees. There are biology students in such places, too, in colleges as well as high schools. But. . .

But I am here. Perhaps, unlike Thoreau, I have only one life to live, and cannot spare time searching for others.

Thursday

"Tell them why we can't understand the book you wrote!"

HATS OFF!

Some of our policies are difficult to defend. We say, for example, that students must wear shoes in the building. Although we are not really afraid of hookworm in central Iowa, our explanation generally is that bare feet are a health problem.

And some of us still ask students to remove their hats in classrooms. I've wondered what I would say if Paul M. asked why he shouldn't wear his striped railroader's cap in biology class. The answer occurred to me today. I'll reply, "I don't know!" and turn back to watering the geraniums. That should settle the issue.

Describing an experiment by which he could determine whether ultra-violet light helps prevent tooth decay, Jeff O. started with this: "Get a group of hamster and divide them in half."

And Mike R. on an experiment to determine whether smoking causes lung cancer: ". . .Then to prove it you got to take some people and split them. The first half, no smoking. The next half has to smoke. Then if someone get cancer you got a good idea if it is smoking or not smoking."

AMES HIGH'S FLIES

In late spring, we search our town for spots where flies might breed. In October, we check on the effectiveness of the summer's control program. And in winter, we breed our own flies in the laboratory.

NSTA is sponsoring another contest to promote creative teaching. Last year I submitted excerpts from this log, and the entry was honorably mentioned. For this year, I am considering our flies. If I talk to myself on these pages, I'll be better prepared.

Thanks to Harold Gunderson, extension entomologist, for the inspiration! Several years ago, the mayor asked me to serve on his SCAT Committee, and I was assigned to Gunderson's group on pests. Our target, along with mosquitoes and rats, was the housefly.

Although the committee's meetings are over, the work isn't, and my biology classes are involved. By the nature of the seasons, the students first check on the effectiveness of last year's campaign. Late in the spring, they begin their own offensive.

In May, the kids work in teams of two, in an assigned area. If they behave as planned, they introduce themselves to the resident, explain the mayor's clean-up project briefly, and ask permission to inspect the yard and alley. Equipped with a questionnaire, they search for conditions that make Ames an attractive place for pests. They look especially for breeding sites: open garbage cans where fly maggots are developing; bird baths infested with mosquito larvae; piles of lumber or trash where rats dig their tunnels and leave droppings as proof of presence.

Residents made aware of the publicity in the Ames Daily Tribune and on radio and television, are generally cooperative. Those taken by surprise are often skeptical. Some of the comments from students: "She kept asking what we were selling. When we insisted we weren't selling anything, she closed the door." "They said to come back tomorrow. They wanted to call the Chamber of Commerce first." "He followed us around the yard with a fly swatter telling us he didn't allow flies on his property."

The survey findings provide ammunition for further publicity. Residents are urged to use fly-proof garbage cans and to bury their compost beyond reach of flies, to drain the birdbath once a week, and to eliminate cover for rat dens.

In the October follow-up, each student is assigned a section of the telephone directory and is asked to call every fifth residence in the alphabetical sequence. The calling is done in the evening, and each student is expected to make ten successful interviews. A single survey sheet lists the appropriate questions and provides space for all ten responses: Are you participating in the campaign?...How did this summer's fly problem compare with last summer's? How about mosquitoes? or Rats?...There are also questions about specific means of control, such as placing paradichlorobenzene crystals in the garbage can. And along with questions about their own performance in the campaign, residents are asked for opinions about aerial spraying for mosquito control.

As in the spring survey, students enjoy sharing stories about their calls: "A little kid answered and said I'd have to call later, his dad had to go to the bathroom.". . ."It's best not to mention SCAT right away. They think you're kidding and hang up." . . ."She said their apples were wormy this year, and what could they do about it?" Word soon leaked around school that we were breeding flies in the bio lab. Breeding flies? How stupid can you get?

I think we do a decent job of leashing our pets, but I still get a little flak. They aren't all ours, either. When windows are open in fall and spring, every insect that comes in is identified as an Ames High fly.

Though our purpose at first was merely to observe the life cycle, we have found that the housefly is a versatile lab animal. Aside from experimental uses, it is a convenient food for other inhabitants of the lab. Our Squaw Creek minnows, for example, thrive on fly larvae. Our ant lions and Cuban "chameleon" are doing well on the adults.

Tips on rearing came from Paul Dahm, who uses a lot of flies in his work on the chemistry of insecticides. I suppose garbage would be a reasonable growth medium, but the standardized diet from Ralston Purina is easier to explain. It is a mixture of oat hulls and alfalfa meal. Moistened with a water suspension of non-diastatic malt and yeast, it ferments, keeping the temperature at the proper level, and releasing a distinctive odor.

Even though the odor is not as unpleasant as garbage, we operate on a small scale. The cages are peanut butter jars with perforated lids. A layer of the souped-up cereal goes on the bottom, then a wad of cotton dampened with a powdered milk suspension (twice as much water as the directions indicate). Adult flies feed on the milk, and females deposit eggs among the cotton fibers. The larvae move down into their food and chomp their way toward pupation.

When the larvae are five or six days old, we remove the cotton and add an inch of sand over the food. Before pupating, the larvae move up into the sand. Then we can easily pour the sand into a tray and sort out the pupae. We drop some of them into the mossy area of the terrarium. When the adults emerge, the lizard will be there to greet them. Others, in petri dishes on a layer of moist sand, are available for observing the emergence.

Only the most torpid sophomore can watch this final phase of metamorphosis without excitement and wonder. The front of the fly's head bulges enormously. A sac-like protrusion, the ptilinum, expands and contracts with fluid pressure. In addition to splitting the pupa case, it normally helps the fly escape from the material that nourished it through its larval stages. The bulge then disappears, and the fly looks like a fly.

Last winter we were using a dish of snow to slow down a group of larvae for my camera. Someone asked whether the larvae would recover from this temporary chilling. They did. But we wondered about the limits of this tolerance. As a first test, we placed five larvae (in their third instar - between their third and fourth molting) in a dish of packed snow on the ledge outside the lab window. Nine hours later, we found that the snow had melted, and the larvae were at the bottom of the dish, in water just above the freezing point. After 15 minutes at room temperature, four of the larvae were again active.

Would pupae be more cold-resistant than larvae? It would be good if we could keep our culture going by putting pupae in cold storage, bringing them out to resume development when they were needed. Would this really work?

Someone asked the question during a class discussion. A work group, one member from each of my five sections, investigated. They placed 20 pupae in each of four test tubes with a little damp sand, then stored them in the cafeteria's refrigerator at 36 degrees F. At weekly intervals, they brought one tube back to the lab, where the temperature averaged about 73 degrees F. They placed the pupae on a circle of damp blotting paper in a petri dish. Of the 20 that were chilled for one week, 14 emerged. Of the two-week hatch, nine emerged, and of the three-week batch, twelve emerged. The four-week batch was accidentally left uncovered for about 36 hours. None emerged, but we don't know whether dehydration was responsible.

That's one of many unanswered questions. Any volunteers?

Adams Versus Schmidt

Verna Schmidt has been principal of Central Junior High for many years, and her hearing problem has not made her more flexible. She is unwilling to bend her schedule to mesh with the Senior High, right across the street. A bit of bending is essential because several of the teachers have duties at both levels. This has been an endless problem for Herb. Logic is no help.

At bridge club, Herb was recounting the latest battle. He had been saving a choice strategy that would devastate her defenses. Sensing that this was the time, he allowed her to talk along without interruption. (I suspect this is not an easy restraint for a principal at any level.) By the time Verna's offensive had run down, Herb's adrenalin had run up, and he pressed the trigger. His arguments must have issued like volleys from a machine gun, with no pauses for reloading.

His ammunition expended, Herb rose from his chair. Miss Schmidt rose, too, without visible wounds. She removed her hearing aid and tapped it with her forefinger.

"Mr. Adams," she said, "some day you'll have to tell me what you've been saying. My battery must be dead."

David tells us that he owes his proficiency in wrestling mainly to the principal of Central Junior High. Miss Schmidt had a habit of putting her arm around him and reminding him that his father was a teacher. He must set a good example for the family.

"That's what did it," Dave explained, "twisting loose and making an escape."

I Wanted to Cheer

Even before Susan had stopped speaking, I wanted to stand up and cheer.

We were discussing the blood circuits that have evolved in vertebrates. The fish, with its single pump, is disadvantaged. The ventricle must force blood through two capillary systems, the gill membranes, then the remainder of the body, before it returns to the heart for another push. There's a loss of pressure.

Birds and mammals have the advantage of two pumps. The right ventricle sends blood through the lung system. Then this refreshed blood returns to the heart and is pumped out to the body cells by the left ventricle. More uniform pressure.

Frogs and other amphibians have a design that corrects the pressure problem of the fish, but it has a flaw. With two circuits, one to the lungs and one to the body cells, there is only one pump. Both the expended blood from the body and the recharged blood from the lungs pour into a single ventricle.

The students had examined diagrams of this design. They could see that depleted blood from the body and aerated blood from the lungs appear to mix in the ventricle. If this occurs, some of the corpuscles that are already loaded with oxygen will float back through the lungs, a useless trip. And some of the corpuscles that have already delivered their oxygen will make a useless trip out through the systemic vessels. A pointless waste of energy!

This apparent defect in amphibian design has led to many investigations and controversies. Some zoologists have argued that a peculiar spiral-shaped valve in the aorta effectively separates the two blood streams. But clear-cut evidence has been ellusive. The injection of dyes, followed by X-ray examination, for example, has failed to settle the question.

I asked the students if they could think of any other way of getting evidence. They looked blank. When I was about to answer my own question, Susan spoke up.

"This sounds a little crazy," she said. "But if you had some way of testing the blood in the two circuits, finding how much oxygen it contains, wouldn't that help?"

"How would that help?" I asked.

"Well," she continued, "if both circuits have the same amount of oxygen, the two blood streams must mix together. But if blood going to the body has more oxygen than blood going to the lungs, then the spiral valve must really work."

Susan laughed nervously. "But I guess there's no way of measuring the oxygen inside the vessels. Is there any way?"

I wanted to cheer.

I had planned this discussion as a lead-in to explain what Karl Delong did at the University of California in Berkeley. Susan explained it for me. She was wrong about some of the details, of course. She didn't know that he extracted blood from several vessels with a micro-syringe. And she didn't know the outcome.

Delong found that blood in the carotid arteries (to the head) had a high oxygen content. Arteries delivering blood to the lungs and skin (which helps in the exchange of gases) were low in oxygen. Blood flowing to the muscles and body organs had a medium amount .

It wasn't necessary to ask Susan if she had read Delong's report in Science, the scholarly publication of the AAAS. If she had read the report, she would have been more accurate. She was thinking.

THE MIRROR

On Switch Day, teachers are expected to trade places with students. There were volunteers in only three of my classes, and all moved smoothly - even in the second period, when Bob S. and Paul S. were performing more for my benefit than for the remainder of the class.

They entered from the staff workroom, wearing neckties and lab coats. Reminding us to take notes, they launched into cancer problems with definitions, statistics, symptoms, treatment, and so on. During one of the transitions, Paul unbuttoned his collar and loosened his tie. Later he hopped onto the lab table and sat there continuing his comments.

When Bob took over again, he asked if we saw any connection of malignancy to the lessons we'd been having about DNA. There were no responses, and he added, "Perhaps you should review your notes and think further about it before class time tomorrow!

While Paul was again talking, he slid down from the lab table, cleared his throat, and drank from the hose that we had fixed as a fountain over the sink.

"I've been meaning to remind you," said Paul, "that this fountain is for your convenience. Yesterday when I came in, water had been splattered over the books and papers. If it happens again, I'll remove the hose." Then he tore several sheets of paper into small pieces and passed them down the rows, telling us to close our notebooks. While he was doing this, Bob was writing on the chalkboard:

What are the seven symptoms of cancer?

What are the three types of treatment?

I knew the three treatments but could remember only six of the symptoms. At the end of the period, Bob and Paul stayed behind looking over the quiz papers. I believe mine was the only one they marked. Handing it back to me, Bob explained that they decided not to count the question about treatment; its meaning was not clear. And on the other question, they could hardly consider six of the seven items a satisfactory record.

I said I would try harder.

A Marsh for George Hendrickson

There's now a marsh dedicated to my former professor. The only better memorial that comes to mind is a square mile of woodchuck and cottontail range in southeast Iowa. But each of us sees George Hendrickson in the way he touched our lives. If I had studied the nesting of the blue-winged teal or the migration of the mallard, I would have favored a place where cattails and rushes grow. The marsh is good.

Hendrickson was in charge of my first course in zoology at ISC. One day I knocked a preserved lobster off the table, breaking the vial. Hearing the crash, Hendrickson calmly brought a towel, handed it to me, and looked down at the lobster. In the same relaxed tone he used in talking about its dorsal nerve cord or its gills, he said, "It was already dead."

Zoology 23a was the first in a sequence of three courses that I had elected at a time when I was vacillating between forestry and English, or perhaps between music and vocational education, so I chose a course designed for pre-veterinary medicine. When I reported to Dr. Hendrickson that a conflict prevented getting the third course into my schedule, he remarked, "You'll be back."

I came back. After graduation, I accepted the job teaching biology at Keokuk. Although I had thought of biology only as a hobby, I was soon convinced that I must learn what makes living things alive. After a bit of groping for directions, including a summer fellowship in psychology, I landed a summer job as state park naturalist for the

Conservation Commission. That's when I returned to Hendrickson. I asked about doing some fieldwork for graduate credit.

"Do you have a species in mind?" he asked.

I named several predatory birds. He looked at me quietly, then leaned back in his chair and took the toothpick from his mouth. "Have you considered people?" he asked. "They are an interesting species." Perhaps he knew of my summer in psychology, when I spent most of my time teaching the human species to drive an automobile.

We settled on the woodchuck.

While we talked, I stared at the odd environment of his office, a musty library of mammal skins and wildlife bulletins in the west end of the Science Building's catacombs. Hendrickson noticed my interest. "I like it here," he said. "If I moved upstairs, I would have a telephone. If I had a telephone, I would have to shine my shoes and brush my teeth so they would be as white as the dean's."

Hendrickson's instructions were to locate some dens and watch the animals for a while. "You'll soon have something to work on," he said. Later, when the pattern of my research began to emerge, I asked his advice on ways of catching and tagging the woodchucks. And he replied, "I assume that is part of what you are going to learn."

Some years later, after I had come to Ames High, Hendrickson thought I might want to spend some weekends and the summer with cottontail rabbits. (Did any of his students escape some kind of contact with cottontails?) One day he accompanied me to the field to see some nests that I had located. We were in the Arboretum, southwest of the campus, and it occurred to me that it would be fun to test my professor's insight into the ways of rabbits.

There's a nest within a hundred feet," I said. "If one knew about rabbits, could he predict the location?"

Hendrickson turned. He turned slowly. He yawned and said it was about time for coffee. Then he pointed with the stick that he used for poking around in the cover. "If I were a rabbit," he said, "I would take a spot this side of the old fence row, say about ten steps."

I wondered why he had hired me to find the nests. He was right.

Returning to his house on Arbor Street, we stopped in the garden. He inspected the potato plants, poking the leaves with his stick.

"Beetles," he said, "but not too many. Some people spray, some dust. I'm not sure but what a little leaf-pruning is good for potatoes."

Mrs. Hendrickson had evidently seen us coming. The coffee was ready, and she remarked that George considered hot coffee to be no more dangerous than vitamin supplements.

After recovering from his first heart attack, Hendrickson told me how he had crossed the street down in campus-town. "I go a little slower than cars, you know. I just hold up my hand and stop the traffic."

I never observed this performance, but I suspect that it wasn't very different from the way George Hendrickson advanced on a rabbit nest or a cattail marsh.

Brief Orbit

Boston, 1958. Someone wrote a letter to the Globe advising the editors to retract their stories about a Russian space ship orbiting the earth. This is an outrage, the writer declared. Responsible editors would not perpetuate such nonsense.

To me, a space ship seems just as reasonable as my being in Boston. The Ginn editors asked me to consider writing a biology text, and I am here to confer on sample chapters. Much of our conference has concerned my point of view. These people are cagey. They don't tell me what they want. I am not sure how they distinguish between structural and functional approaches. So I try to be cagey. I simply tell them what I do.

Now a letter from Ginn advises me that although they like the way I write, my viewpoint doesn't meet their needs. They apparently want an encyclopedia of organisms, with all the parts labeled. This, I suspect, is an admission that Holt's Modern Biology is outselling their old Curtis production. I spend too many words on the way biologists work, on why living things are designed as they are.

With a gesture of good will, the editors say that if I wish to place more emphasis on structure, they will be happy to hear from me further. They added that if my convictions are firm, however, I will probably not wish to change directions. They are right. I agree with

Woodrow Wilson of Princeton, who advocated "developing the mind by using it rather than stuffing it."

FIRST AID IN BIOLOGY?

Is first aid a proper study of science? If you are in tune with the times, you say no. Stopping the flow of blood, applying artificial resuscitation, keeping microbes out of a wound - these are technology.

Call it what you will, I teach first aid in biology. I have objected, however, to a common way of testing students about injuries. The usual approach is to ask, what are the dangers of moving a person who has a spinal injury? How would you treat a person who is suffering from heat exhaustion?

You don't find an accident victim labeled concussion or compound fracture or arterial bleeding. You inspect the victim. You treat according to what you observe. In my tests, therefore, I emphasize symptoms, not the names of injuries. Here is an example:

It is hotter than you thought. The portage is longer, the canoe heavier. After fighting your way through the alders into a clearing, your partner sags against a fallen tree. You ask him if he is okay, and he doesn't answer. Then you see that his face is pale. He is sweating heavily. He vomits. You check his pulse and find that it is weak and fast. He is conscious but woozy.

a. What will you do for your friend?

b. What is his condition called?

Checking a set of papers yesterday, I found that Gordon had done a good job on the first part. For the second part, he said, "I suppose you want us to say that it's heat exhaustion, not sun stroke. But I think the question isn't really fair. Most of us don't even have a canoe."

Gordon, if I ever have to carry a canoe through the alder thicket, I hope you'll be there.

THE LEASH

For months, my mind has been on a leash. The restraint began when Herb Adams asked that I serve as chairman of the Steering Committee. Our task was to guide the faculty in a self-evaluation, as required by the North Central Association.

For a week now, my mind has been loose. Even so, there are moments when it remembers the leash. While occupied with the mundane affairs of prairies and raccoons, genetics, and biology tests, it is jerked back to philosophies of learning, integration of the humanities, community resources, and a misplaced report submitted by the math department. And my mind still looks for an answer to the question, if this is a self- evaluation, why am I involved with football coaches and art teachers?

In a way, it was a relief to have the critics descend on us for several days. Teachers from other systems, administrators, university professors, and people from the Department of Public Instruction - they browsed through our classrooms, chatting with teachers and students, deciding whether the grueling diagnosis that we submitted was realistic.

Following a sort of banquet, I moderated some introductions and greetings, during which the chief critic said some nice things to the Board of Education. Written reports will follow. I've forgotten what the nice things were.

NEW WINDOWS

A few years ago, when the school bought an extensive building site north of our home, I dreamed that the new high school would be equipped with a tunnel opening into our basement. And there would be a woodchuck range and a greenhouse attached to my laboratory.

Now the first phase of construction is complete. There is a wing for industrial arts and home economics, another for science. These departments have moved in, along with a few math classes. The remainder of the high school will stay downtown until additional new space is available. Adams will remain at his old office, and

Ritland will be responsible for the new unit until the move is complete.

I have not located the tunnel. Nor the woodchuck range.

The greenhouse is real. It opens from a workroom between the two bio labs, a few steps from my desk. There is also a door from the greenhouse to the outside, and that is an exciting feature of my new base camp. The site was part of a hilly farm, brushy woodland and pasture that had never been plowed. A number of prairie species have survived the grazing of cows and horses. Although a considerable part of the tract was disturbed by bulldozers during the construction, there is much worth saving. If we can keep football fields and parking lots within bounds, there will be 20 to 30 acres left. Botanists from Iowa State have already insisted that we must preserve what is left of the prairie. I have planted three colonies of bees, hoping that squatter's rights will be respected.

My indoor space is room 111, a combination laboratory and classroom. It is at the remote end of the science wing, closer to the prairie than to the administrative offices. There are ample storage units, electrical outlets at each of the 15 student tables, water and gas at the back of the room as well as in the elevated demonstration area at the front.

The platform was too short to accommodate my desk. With a leftover plank, however, I extended the platform, and the desk now butts up against the demonstration unit.

Out the windows, I can see the University's pine grove to the southeast. (The tunnel should pass beneath the corner of the grove.) Eventually, a two-story wing will block this view. I will look into an enclosed courtyard.

Still, I will know that to the west between the prairie and the woods, a little stream trickles past the floodplain into Squaw Creek. This is good.

The Unsure Earth

We have considered buying an updated world atlas. But I'm alarmed by the talk of some of the geologists. They say the continents are floating around on big rafts made of rock. It's part of the theory of plate tectonics. The rafts don't sink because they're floating on rock that is denser than what's on top. It's like a chunk of steel floating on mercury.

If the speculations are correct, the earth's land was once a single immense continent. It is called Pangaea. When Pangaea broke up, the pieces drifted apart on their stone rafts.

One bit of evidence favoring continental drift is the jigsaw idea. If you could push Africa and South America together, they would fit reasonably well, like a puzzle.

The thing that alarms me is that we are apparently still adrift. Africa is headed north, South America northwest. Ames High School, with most of the continent that it sits on, is headed roughly west. A piece of California is shearing off at the San Andreas fault and migrating north. The rate of drift, they say, varies from about one to fifteen centimeters per year.

This is discouraging news. I think we'll delay shopping for an atlas until things settle down.

Tolerating the Intolerable

When a marsh becomes overpopulated with muskrats, these animals become less tolerant of each other. They fight. The weaker members of the community are evicted. The outcasts, living in marginal habitat with exposed den sites and too-shallow water, are killed by marauding minks and horned owls. The muskrat population drops to a tolerable level.

In a fishpond, balance is achieved in a different way. If a pond is capable of producing 100 pounds of bass in a year, and the pond is home for 100 bass, the average gain will be one pound. If there are 200 bass, the average gain will be a half pound. And with 500 competing for the food supply, the gain will be only a few ounces.

The people problem is still different. Our ingenuity has permitted expansion at a malignant rate. Neither war nor famine adequately replaces the role of the mink in the marsh. We quiet our conscience by providing enough food to keep the reproductive rate ahead of the death rate. We tolerate the intolerable.

If we humans would devote as much energy to the development and distribution of good contraceptives as we do to spears and bombs, we might live on a better-balanced planet.

Heart Sounds and Skeletons

When Julian McFarland learned that there was no human skeleton in our lab, and only one stethoscope, he contacted his medical colleagues who had graduated from Ames High. They contributed.

With the resulting fund, we purchased one skeleton and fifteen stethoscopes. And that wasn't the end of the benefits. Dr. McFarland squeezed the Board of Education, and the Board reconsidered our request for additional microscopes. Now every student in a class can work with a personal scope.

As for the stethoscopes, there's a great advantage in having all of the students (in teams of two) working on a problem at the same time. A single scope for a class of sophomores frustrates both the teacher and the students. Sure, some of the same investigations are possible by a touch of the fingers over the radial artery in the wrist. But a stethoscope adds impact.

I recall the alarm of one of my students when we were checking heart rates before and after exercise. She came to me and said there must be something wrong with Clif. He had only 34 beats in a minute. Clif looked healthy enough, but I went over and took his wrist. I expected to find that the count was 68, that the student had counted for only 30 seconds.

Thirty-four thumps per minute. That was it.

"Are you going out for track?" I asked. Clif nodded and grinned. He earned his letter and then some. After starring at Kansas, he went on to the Olympics. Only one runner beat him to the tape in the 400 meter hurdles. Then he was shot down in Viet Nam.

Another student, years later, was less interested in track. When he told his dad, a physician, what we were doing with the stethoscopes, the doctor loaned us a tape recording of abnormal heart sounds. We weren't trying to make little doctors of our sophomores. But after hearing the swishy sounds from leaky valves, the students had more appreciation for a normal heart.

Doctors aren't the only biological specialists who have enriched our courses. Beekeepers, farmers, bird watchers, sanitary engineers, foresters, pest exterminators - a long list of adults have responded when we've asked questions. Their help has even led to things like stethoscopes and skeletons.

I Tried an Analogy

If all else fails, I look for an analogy. That's how it was when I was trying to help students appreciate why the release of energy in respiration is such a long sequence of reactions, and why some of the reactions appear to be headed in the wrong direction. .

"How come the molecules are getting bigger?" someone asked. "I thought you said respiration breaks glucose down."

He was referring to the reactions in which pyruvic acid is broken down to acetic acid, then rebuilt into much larger molecules of citric acid. How come?

Well, I explained, to understand why the cell doesn't oxidize acetic acid more directly, you would need a better background in biochemistry. So would I. Let's just agree that the cell can't go around making enzymes for every conceivable reaction. It gets along with what it has, transferring energy gradually, in amounts that the cell can cope with. And you should know, too, that some of the part-way molecules in respiration, like acetic acid, are useful in other ways besides energy transfer. They are starting points for other valuable substances.

"Okay, but I'd think there would be a simpler way of getting energy out of glucose."

There is. Light a match to it. But let's back up to the original question: why is the respiration route so indirect, and why does it involve some back-tracking? If you drive across the continent in a car, you don't go in a straight line. You follow existing highways. In a mountain range, you may have to drive uphill for a time when your destination is really downhill. It's easier to stay on the road than to drive in a straight line. Chemical reactions are like that. They take the easy route.

There! Why hadn't I said so in the first place?

But analogies can get out of hand. I remember an example from my chemistry teaching days. Trying to clarify the affinity of certain elements for each other in a reaction between two substances, I asked the students to think about the pairings that occur on the dance floor after a basketball game. A couple may seem happily united until another possible partner shows up. After pursuing this idea for a while, I had the feeling that we weren't making much progress. And I was sure we weren't when one of the boys asked, "Mr. Trump, could we try going back to atoms and molecules?"

That was years ago. Just a few days ago, we were contrasting the endocrine and nervous systems. Which one delivers messages more accurately? The nervous system - good. It's like a telephone exchange, the books often say. You dial a number and the message goes in the right direction, and only in the right direction. An endocrine message is more like broadcasting an announcement over a loud speaker. Although it may be intended for one person, it reaches a lot of ears. Generally, however, only the target person responds.

That had a better effect than the dance floor idea. At least as I cast the message broadly from the front of the lab, several of the targets looked as though they'd been hit!

How About Some Organic Water?

A dealer came around last week suggesting that my students could earn a pile of money for our pond project selling vitamins - organic vitamins. When I asked how an organic vitamin is different, he wasn't sure. But with all this ecology," he added, "organic things are going over big."

And now our paper reports that the Federal Trade Commission is clamping down on a firm in New York that is advertising organic sugar. The firm claims that the sugar is "raw" and loaded with nutrition.

It's odd how a simple term like organic can go wild when it escapes from the vocabulary of science. For that matter, a simple term can go wild when it escapes into the vocabulary of science. Well, the organic confusion started in 1807 with the Swede,

Berzelius. He observed that some substances were associated only with life; they were organic. Others did not require a life origin; they were inorganic. All organic substances, it turned out, contained carbon. Most inorganic substances did not.

Twenty years after the birth of organic, however, Friedrich Wohler found that he could concoct urea from an inorganic salt, ammonium cyanate. Today chemists can build complex molecules in seemingly endless variety. And you can take your choice of two reasons for calling such molecules organic. They contain carbon, and they are made by organic chemists - live chemists.

I suppose the term really escaped when gardeners got the notion atoms of elements such as phosphorus and potassium that come from manure are better than the same kinds of atoms that are wrapped in plastic. Organic vitamins were bound to follow.

When I asked the salesman where his vitamins were produced, he was happy to say that the factory was out in Omaha. Has anyone thought of bottling demineralized water and selling it as the good old organic stuff?

"Unatomical Speciment"

It's been a good many months since we ordered an articulated human skeleton from a firm in Calcutta. Frank Howell, Secretary of the School Board, disapproved. He feared that the original owner of the skeleton had died of bubonic plague. It would be a source of infection for all who touched it. As weeks went by, I began to fear that Frank had reached the immigration authorities and that the skeleton was being held on Ellis Island until it could prove itself in good health. Here is an abbreviated review of the correspondence:

From Reknas Limited, Calcutta, October 19: "Dear Professor Trump, we have your cheque for $100. We are taking up the matter in hand and shall write you again after dispatch."

From Reknas again, November 16: "We have the pleasure to inform you that the order for an articulated human skeleton has been executed." (Take comfort, Frank; perhaps the owner was healthy at the time of execution.)

From Norton, Lilly & Co., New York, December 18: "Ocean freight charges payable, City of Manchester, pkg. unatomical specimen, wt. 100, prepaid."

To Norton Lilly & Co., February 12: "It has not arrived. What next?"

From Norton Lilly & Co., February 19: "The pkg. of anatomic speciments of which you inquire was landed in New York from S/S City of Manchester. As the pkg. was not permitted in the customary manner, the Custom Inspector ordered same into General Order storage, Lehigh Warehouse Co., Newark, N.J. We suggest you appoint a Customs Broker here in N.Y. to arrange early clearance of an articulated human skeleton."

To Customs Inspector, Port of N.Y., February 25: "According to Norton, Lilly &...S/S City of Manchester...anatomical specimens commonly imported...reason for delaying shipment...please inform us."

From Customs, April 7: "Dear Mr. Truml, The permit for release and your check to Lehigh Warehouse were forwarded with instructions..."

From Phoenix Assurance Co. Limited, 2 Hare Street, Calcutta (not dated): "...Subject to terms and conditions...Item 12. Warranted free of capture, seizure, arrest, restraint or detainment and the consequences thereof..." (It escaped? Even after execution?)

Because we were running low on postage at Ames High, I decided to cease and desist for a time - that is, to take the matter in hand out of hand. Good decision. A week later 1 pkg. arrived at the receiving room of the Ames Community School District; the Secretary of the Board ordered that it be delivered promptly to room 111, Ames High School.

A few days later, a letter arrived from Norton, Lilly & Co., Steamship Agents and Operators, Regular services: Australia-New Zealand, India-Pakistan-Ceylon, Philippines-Japan-Straits Settlements, Egypt-Red Sea-Persian Gulf, South and East Africa, Brazil-Uruguay-Argentina: "This is to inform you that one pkg. unatomical speciment was landed by the above vessel in New York..."

Despite the ordeal of such a journey, I find the speciment to be in tolerably good health.

Ames High's Mountains

Not satisfied with the work of glaciers, water, and wind, the developers flattened the prairie contours in front of Ames High School. Then they hauled soil and made three abrupt mounds that obscure visibility at some of the parking intersections. I understand that students have named the highest peak Mt. Hetzel, for our superintendent. The second is Mt. Adams, for the principal. And then comes Mt. Ritland, for the vice-principal who deals with most of the discipline problems.

The CabelLegacy

Registration of sophomores is an administrative task. Since I do not administer, this day was mainly free for restoring some of the clutter that was hidden when schools closed last June. I was pulling expired notes from the bulletin board when a gaunt boy ambled in and asked, "Do you know anything about grasses?" His question was more like a demand than a request.

"I know bluegrass and a few of the weedy ones, along with most of the crops-and I know the natives that are out there on the prairie. Why do you ask?"

He responded by waving a greenish-brown bouquet toward me and said, "I don't know this one." Neither did I. But R.W. Pohl's, How to Know the Grasses, was on the shelf behind me. "This should help," I said, not really believing my words. Grasses are difficult.

For the next ten minutes, the boy sat at the corner lab table, turning pages while I arranged some fresh items on the bulletin board. Then he came forward and announced, "It's not in the book."

What he meant, I assumed, was that none of the drawings clearly matched his specimen. He needed to know more about culms and glumes and ligules. "We are having a public tour of the prairie at four o'clock on Labor Day," I said. "Dr. Pohl will be there. Can you come?" He nodded, and I added, "Tell me your name. I'll see that you have a chance to talk with him."

"Carl Craine. I'll be in your honors biology class."

On the following Monday, Carl Cabelwas there at the prairie gate with his bouquet. When Pohl arrived, I introduced the two and said, "Carl would like your help on a grass."

The boy flourished his bouquet and said, "It's not in your book."

"Well, let's see," Pohl replied. He took the specimen, examined the junctures of the leaves and stem, then opened the hand lens that always hangs from his neck on a cord. After inspecting the fruiting heads, he looked at Carl, narrowing his eyes and said, "It's not in the book! Where did you find this grass?"

"I'm not prepared to say right now," Carl replied. "There is only a small patch."

Occupied with the sign-up sheet for visitors, I did not hear the remainder of the conversation. But minutes later Pohl motioned me to the edge of the group. "Who is that boy?" he blurted. "That damn kid knows more about grasses than some of my graduate students!"

"All I can tell you now is that he lives west of town on a farm and that he will be in my biology class."

"West of town - west of Ames? He won't tell me where he found that grass. It's not in any book!"

In the months that followed, I found that I could give other students in my honors class a chance to speak if I ignored Carl Craine's offer.

And if I had trouble identifying a plant, native or otherwise, I could depend on Carl Cabelfor help. I saw him wandering around the prairie during school hours, looking as though the wind might blow him into the pine grove. Because of fragile bones, he was excused from physical education, and Ray Smalling thought the prairie hills would substitute. I learned, too, that Carl had visited further with Dr. Pohl, and that eventually he revealed the location of the errant grass, along the Northwestern railway, between Ames and Boone. Pohl's appraisal: "Cabelwill make a good botanist if you can domesticate him."

Molecules of Memory

Blain stopped at my desk yesterday, blocking the way for those arriving on time. He wondered if I had heard about the man with the memory problem. I should have said, "I can't remember, Blain. Tell me." (I'm good with replies like that. They come a day late.) But Blain didn't really want a reply. He launched into the problem of the man who couldn't remember his telephone number, his wedding anniversary, his social security number, even where he had parked his car. While the man was grumbling about his problem to his psychologist, the telephone rang. The doctor answered, confirmed an appointment, and hung up. Turning back to the patient, he asked, "When did this trouble begin?"

And the patient replied, "What trouble?"

When Blain finished the story, I should have said, "Blain, do you happen to remember that your lab reports are due today?" (I just now thought of that.)

As it happened, though, we were currently studying the human nervous system. Blain was tuned in better than usual. The previous day's class had ended with a question. Dexter had apparently been reading about planarians. "Is it true," he asked, "that if you teach one of those little worms to go through a maze by turning left every time, and then you kill the worm and grind it up and feed it to another worm – Is it true that the second worm will keep turning left and go through the maze without any training?

I looked around the room to see if anyone had been ignited by this, and Judy volunteered. She said she had read the same thing, in an old issue of Science World. Supported by Judy's comment, Dexter moved into the question that he'd been leading up to. "Okay," he said, "then maybe you could train an animal to do something, and you could eat the animal's brain. Then you could do what the animal was trained to do. Is that possible?"

That's when the bell rang. My main thought - wouldn't it be great if physics were taught in the ninth grade, general chemistry in the tenth, and biochemistry in the eleventh. Then biology. We'd be ready for those molecules of memory.

Lacking biochemistry, we disposed of Dexter's question hurriedly today and went on to the synapse. After the students had read about the synapse, I thought they might appreciate one of the organizing principles of the nerve system - why impulses don't travel in the wrong direction over a neuron route. It's because of the gap between the axon endings of one neuron and the dendrites of the next. The impulse moves across the gap only when a neurohormone, secreted by the axon tip, diffuses across that microscopic gap and activates the dendrite. No such chemical is produced by the tips of dendrites. Therefore, an impulse cannot move from a dendrite to an axon. The synapse acts as a one-way valve.

The model that Kristie and Dan made last year helps explain the nature of the synapse. They used white toothpaste on a sheet of black cardboard for an enormously enlarged pattern in relief. It shows several synapse "connections" between two neurons. The raised layer of toothpaste for a cell body, with a darkened nucleus, is convincing. In fact, I've looked at the model so often that now when I see an electron micrograph of the real thing, I have the impression that toothpaste is better.

Well, not a bad day. Most of the lab reports came in. And at the end of the second period I had time to ask Blain if he'd like to tell his story for the class. Blain started to rise from his chair. Then with a teasing glint, he sagged back and said, "What story?"

One Stuffed Eagle

There's only one stuffed eagle in my lab. It is perched above the east storage cabinet, its dehydrated talons clutching a knobby limb. Its wings are raised as though it is ready to swoop, and its glassy eyes glare down at my sophomores.

After a few days, the sophomores stop glaring back.

Once each year, I bring the eagle to a lower elevation, atop the demonstration table. It's good conversation for a lean day.

Student: Why doesn't it have feathers on its head?

Student's lab partner: It's a bald eagle, stupid.

Another student: Bald eagles aren't bald, stupid. They have white heads.

Still another: But its head really is bald - I mean bare. So what kind of eagle is it?

Eventually, the hasslers look in my direction. It's a bald golden eagle, I explain, feeling a surge of authority. It's bald because some little birds picked feathers off its head. Tom P. brought it in during my first year at Ames High. He said his grandfather shot it in western Nebraska 50 years before. Tom said we could keep it for a while.

The while stretched into enough years that the bird must now be at least 80. It's health began to fail, I believe, when the drama department used it for a prop in a school play. I considered throwing it out, but kept it because I enjoyed explaining its feathered legs, feathered all the way down to its feet. Bald eagles have bare tarsi. Artists who draw our national bird often give it the white head of a baldie, but the legs of a goldie. They like to draw feathers.

From the rear row: You said little birds picked its head. Why did they pick its head?

This question shows that I am still in charge, so I explain how little birds have a habit of picking on big ones, especially during the nesting season, and especially if the big ones are stuffed predators. Well, we took it out one spring day and mounted it in a haw tree so we could observe the songbirds that came to heckle it. They picked at its head. But remember, it was an old eagle. It didn't have many feathers to pick.

If there is a little more time, I explain how I got up early one winter morning and tied the eagle to an oak in Brookside Park. It looked remarkably like an eagle. And it happened to be along the route where my classes would be spotting birds in preparation for helping with the Audubon "census."

I guided the group in a circle around the oak, about 50 yards away. We saw a chickadee. Moving a little closer, we saw a nuthatch. Then a brown creeper and several juncos. I was afraid to steer the students any closer, the eagle might fall off and hit someone on the head. Stalling, I suggested that we check on our lists so far. Looking at their notes, the students named about a dozen species.

"Any others?" I probed.

"Well," said one of the girls casually, "there's an old eagle up in that little tree. But what's it doing down here in the park?"

That's another reason why I can't spare our one stuffed eagle. Some day we might have another winter field trip.

"An Excess of Which is Not Needed"

From Joel K.'s notebook: "To give off an excess of which is not needed." It was an isolated entry, in a section that I believe related to our study of protozoans. Was Joel referring to the contractile vacuole of a paramecium? The potato surplus in Idaho? The carbon deposit on the valves of his souped-up Ford?...

Whatever the intent of the entry, it inspired another attack on the storage-of-information problem. I duplicated a guide sheet and asked each student to study his or her notebook and score it as indicated by these questions:

Are your notes accurate?

Does your notebook show regular use? Is it complete?

Are your drawings labeled well enough to be useful in review?

How is the organization?

Is your notebook in good physical condition?

Is it neat?

Does it contain a complete record of your test scores?

Each question except for the last one is illustrated by brief comments and examples. The maximum score on each of the first four questions is 20 points. Neatness counts ten; physical condition and test record get five each.

After allowing students time to examine the scoring sheet, I explained that I would compare their self-scores with those I had already compiled, but that neither score would be entered in my grade book. This was a somber experience for many of the students, and for several days they were more careful in their records. For several days!

Now if I were a psychologist, I would list all the students who rate their notebooks higher than I do, another list of those who are more critical of their work than I am. Then I would examine a number of other variables and see how they correlate with the two groupings. Do the athletes fit mostly in one of the two groups? Which one? How do those with high scores in standardized reading tests compare with those whose reading scores are low? Is there any relationship to the educational level of the parents? Are those who drive their own car to school more critical or less critical than those who walk or pedal a bike?

This would be interesting. But if I were to get involved with correlations and analysis of variance, I would not have time to watch woodchucks or check my students' notebooks.

Diving in the Lab

Teachers who want all of their lab investigations to come out with clear-cut conclusions should not mess with the diving reflex. We have tried it every year since Scholander's article appeared in Scientific American. The results have been frustrating-and exciting.

I like it that way, at least part of the time.

When seals and other diving animals submerge, their pulse rates drop. Part of their blood circuits are largely closed, with only the most vital organs getting full service. This is one reason why such animals can stay under water so long. Their oxygen supply is rationed to the most essential tissues.

To a degree, many other mammals and birds are equipped with this safety device, Homo sapiens included. Scholander calls it "the master switch of life."

Without preparatory reading or discussion with the class, I have asked students how their heart rates would respond to having their faces under water. To test their answers, we filled a plastic tub with water.

Four-member teams performed the test - one timekeeper, two pulse checkers, and the diver. Adequate controls are demanding. The subject's pulse must be determined both before and after, as well as during the dive. For all three counts, the diver should be in the same body position. Ideally, immersion should be the only external factor affecting heart rate.

The data we collected during those initial tests were somewhat irregular. For many of the divers there was a convincing drop in pulse rate. For some there was no change. And for others there was a slight rise. Were these differences due to variations in the basic physiology of the divers? Were the pulse checkers having trouble? Had we overlooked a factor that should have been controlled?

The follow-up discussions were lively. I began to believe that the kids were gaining more from attempting to explain the contradictory data than from seeing the utility of the reflex.

With greater confidence in the students, I decided to scuttle the duplicated instructions. I now tell the students that they are to work in five-member teams to investigate the effect of the tub dive on heart rate. They have class time for planning the procedure before starting the tests. Commuting among the teams, I listen to the comments but offer advice only when asked.

Individual students would surely blunder into more difficulties for lack of reasonable controls. As groups, however, they do as well as when they are provided with step-by-step instructions. And I suspect the critiques by their peers have more effect than anything I would say. A good deal of this pressure comes after all tests are complete, when the teams are reporting to the class.

How do we explain the variable results? The most interesting hypothesis is that psychological factors such as tension, fright, and excitement may work against the reflex, causing a rise in heart rate. If so, this may illustrate a defect in the functioning of the human nervous system. If may be a problem of intelligence: recognizing the danger of being under water may deprive some humans of an adaptation that would otherwise help them survive.

"A good teacher is someone who can understand those not very good at explaining and explain it to those not very good at understanding." I don't know who said it. It would be a discouraging thought, however, if John James Audubon hadn't said that the woods would be very silent if no birds sang except those that sing best.

As a help in getting to know my students, I ask them to write on a filing card some bits of information about their interests and plans. My final question is, "what grade do you expect to earn in this course?" This helps, especially when the replies are as personal as two that I received in a chemistry class. One said, "I hope to get an A and if I enjoy it enough I think I can if I try." The other, "That is a difficult question. I study quite hard and am striving for a good grade, but chemistry to a certain extent depends on the individual. I expect a grade to range anywhere from B to C. Maybe my mental capacity will consume more whereas it may be higher or it could be vice-versa, but I think it will be average."

Jan tells me she hopes to be a veterinarian. The note that I found on my desk suggests that her animal - empathy should be an asset: "Mr. Trump, I took the kitty home so it can have some real cat food and won't have to be alone in a big dark laboratory with a lot of awful noises. Then I will try to find a solution to its home."

Perhaps I Was Lost!

This is how it was at 8:15 Thursday morning: Greg was leaning toward Tricia. Tricia was leaning toward Al. Terry and Ben were flipping through a catalog of sparkplugs and mufflers. Edie was staring north, Stephen south. Blain was doubling another stick of gum into his mouth. Ellen was pulling her long pale hair forward, separating the strands with her ballpoint pen. And B. John was holding his pencil at an expectant angle over his notebook.

It was like any other day just after I'd hung the attendance report outside the door.

On any other day, however, my next move would be to walk out in front of my class, hoping I would be seen. Then I would call for attention, hoping I would be heard.

This would be a signal to Denise. She would suddenly have an urgent message for Barbara, three seats away. Paul would find that his pencil needed sharpening. Anne would discover she'd brought her geometry notes instead of her biology report. And Carl would have a coughing seizure. After which B. John would turn and emit an impatient "Shhh... !"

But on Thursday, I didn't ask for attention. I returned to my desk and sat down.

This wasn't planned. It was an impulse. What would happen, I wondered, if I just sat there reading, ignoring 27 sophomores?

The answer came in the next seven minutes.

I turned pages without seeing words. Trying to relax, I stole glances around the room. At 8:17, several other boys had gathered around the automotive catalog. Paul was there, leaning on Terry, disagreeing about carburetor adjustments.

Once when I called Paul's home to see if I could stir action on his lab reports, his mother said she couldn't understand why he would do this to her.

"I even bought him a car," she explained. "I told him he couldn't take it out of the drive till he was old enough for a license, but he promised to work harder if he had a car of his own. He sits in it and races the engine. I can't understand that boy!"

Blain, 50 pounds overweight, was slipping down in his chair, looking at the ceiling and chewing at an easy pace. Tricia had failed to reach Al and was now chatting with Greg. And Edie was marking a paper written in longhand. I wondered if it was another letter to the editor. The last one asked why school couldn't be relevant.

One of our few dress regulations was that students must wear shoes. While Edie was testing the pH of an enzyme solution, I noticed her bare feet protruding from beneath the lab table. I was about to water the seedlings at the window, and it would be easy to pour a little water on Edie's toes.

I poured a little water on her toes.

Without a word, or even a look of surprise, she slipped her sandals on, got a paper towel from the dispenser, and wiped up the water I'd spilled.

By 8:19, the volume of chatter had doubled. B. John, however, had tuned out his environment and was working geometry problems.

One day I asked B. John what the B was for. Although he was generally called John, he always signed his papers "B. John." I was curious. Perhaps it would help me understand his dedicated attack on his assignments. "It's part of my name," he said. And I thanked him.

I turned another page, thinking how unimportant names should be. I should know my students by how they think and what they do, not by the names they write on exam papers. Yet there were at least a half dozen of those 27 whom I knew only by name.

"Why are you always trying to make me ask questions?" said Mike one day after class. "I listen. I know what's going on!" We'd been discussing the possibility that medical science reduces the fitness of humans; by cutting the death rate of people with hereditary defects, doctors enable defective parents to produce defective children. My reply was that when a student asks a question, I have some idea what's going on independently in his head. Merely answering my questions isn't so revealing.

Too often students say what they think the teacher wants them to say.

8:20. The volume leveled off. Several students had turned their chairs so they could talk more easily with neighbors behind them. Blain had oozed down onto his lumbar section. Tricia and Greg appeared to be working on a crossword puzzle in Science World.

Tricia was the kind of student who could erect an impenetrable shield between me and her mind. At the end of a class I would think: six more years to retirement-perhaps I could still become a plumber's apprentice. But on other days she could remove the shield. She could come up with flashes of insight that made me regret that I had only six more years in which to watch sophomores for signs of thought.

At 8:21, I was convinced this would go on till time for the next class. The article on the ecology of the camel wasn't making sense either. I was hearing snatches of conversation: basketball, membership on the drug committee, play rehearsal, the book report for English, why the student council doesn't do something about something. Mike was now wandering from one table to another, thumping fellow students on the head. And Denise had an audience of two boys and one girl.

"Denise!" I once challenged. But she broke in before I could start a sentence. "I know," she said, "I talk too much. My counselor says I'm a compulsive talker. That means I talk too much. But you know, Mr. Trump, it's Barb's fault, too. She listens too much."

"You mean she's a compulsive listener?"

"Maybe that's the trouble," Denise concluded. "I'll tell her not to listen so much."

But I wasn't prepared for what happened next. As though mysterious signals were spreading through the class, one after another of the little clusters of students fell silent. Within a few seconds, there was a hush that caught Denise in the middle of an exclamation point. I'm not sure what I had expected, but this turn of events caught me off guard. How do you break into this kind of silence? I should say something profound. Something like: "If you always wait to be told when to go to work, you'll always be working for someone else." No, this wouldn't do, everyone works for someone else..."You have 37 minutes left for a full period's work."...Well, that idea had come up too many times in the past few months.

So I pretended to read, and I could see that students were glancing furtively in my direction. Blain hoisted himself up. Ben closed his catalog. Paul moved to his own chair.

Still searching for a transition, I stood up. And while I was trying to generate a penetrating expression, Paul broke in:

"We wondered where you were, Mr. Trump. We thought you were lost!"

I decided to settle for that. Perhaps I was lost!

How to Explain an Explanation

Finding a better way can be a humbling discovery. When I blunder onto an obvious improvement, I am puzzled why it took so long.

Years ago, I concluded that it was better for students to know how a muscle functions than to know its name. Eventually I found that it was better to have students do with their muscles rather than talk about them.

We used a big wall chart showing a muscular male who had been stripped of his skin and one of his appendages. He was pictured from both front and rear, in an active pose with one arm flexed partway overhead. I marked about eight muscles with code letters and wrote their names on the chalkboard.

There were two approaches. I could point to a muscle and ask what movement contraction would accomplish. Or I could describe a motion, swinging a tennis racket, for example, and ask what muscles were doing the work. Students were expected to know that a skeletal muscle can move a bone only if it is anchored on both sides of the joint. They were expected to know also that a muscle works by pulling, not by pushing.

When quizzed with the wall chart, students often explored their own muscles by touch while imitating a motion that I described. This doing with their muscles was student initiative. I cannot claim credit. I merely encouraged it.

Then came the day when I thought my sophomores should probe a little deeper. Why do the abdominal muscles contract when you do pull-ups on a bar or climb a rope? The function in sit-ups and leg-lifts is obvious. But why should these muscles work when there is no forced movement at the hip region?

My explanation was clear, to me. But even if the students could repeat what I had said, there was a fuzzy look in their eyes. I remember what Leonard said last year. He was a bit slow, and he often bugged me for a better answer. After my explanation, Leonard said, "Mr. Trump, I wish you would explain it!"

Perhaps Len's plea is responsible for my belated "discovery."

Two tall boys held a six foot length of steel pipe overhead, providing a chinning bar. Another shorter boy, stripped to the waist, did a few pull-ups, facing the class.

What joints were moving? Mainly the elbows and shoulders. The muscles working? Mainly the biceps for the elbow joint, the pectorals, latissimus dorsi for the shoulders. It was obvious, seeing Brad's lean belly, however, that his abdominal muscles were working, too.

My explanation may have been no clearer than last year's. Visible muscles may have made the difference. Instead of using the chart, I pointed to Brad's sternum, the breast bone. This is the lower anchor for the pectorals. The sternum, in turn, is supported by ligament attachments to the ribs and clavicles. These connections, however, do not make a firm anchor. They get help when the abdominal muscles pull downward. Then the sternum is not yanked out of position when the pectorals contract.

And just now I've thought of an analogy.

Suppose you are pitching a tent, stretching out a rope and tying it to a stake. The tent is pulling on the stake. Suppose you find that the stake is not holding firmly. You tie a second rope to the stake, stretch it back of the first one and tie it to a second stake.

Brad's abdominal muscles, tied to the bones of the pelvic region below and to the sternum above, are like that second rope.

Next year when we come to muscles, I'll discover a way of supporting the chinning bar between the top of the storage cabinet and a temporary scaffold. If it needs more support, I'll try a tent stake.

SCHMIDT WINS BIG TOMATO

Room 111, Ames High School. The tall tomato reached the ceiling yesterday at 2:23 p.m. Dan Schmidt was declared winner of the contest.

When the seed was planted last September, each biology student in 111 registered a guess when the plant would outgrow the room. It did so yesterday, having grown 1.2 millimeters per day.

Interviewed in west hall, Schmidt declined to reveal how he knew so much about tomato culture at the very beginning of the biology course. Later, in east hall, when asked if he used a computer in his calculations, he replied, "I don't think so."

Instructor for the classes in 111, Richard Trump, explained that although this is the first year for the contest, Tall Tomato Day is expected to become a tradition at Ames High.

Schmidt prize, awarded at the end of the seventh period, was a partly pink tomato, almost as large as a golf ball.

DRIFTING CONTINENTS, WRINKLED APPLES, SIGHTLESS SALAMANDERS

How one thing leads to another: It's no wonder I know so little about the home range of the woodchuck or the flora of the Ames High courtyard. I was thinking about the upsurge of interest in plate tectonics. If the continents have really broken apart from a huge primal land mass, the theory should help explain the present distribution of life on the earth. Here is the devious route of my thoughts.

Drifting continents? The idea is probably much older than I realize. Was there any hint of it in Shaler's <u>Aspects of the Earth</u>? I read the book 40 years ago, at a time when I thought I might understand God if I knew a little about astronomy. And if I am to deal with the universe, I'd best start with my own planet.

About half way between then and now I was consulting with Charles Gwynne one day about the identity of a fossil that I found in the Des Moines River valley. For some reason, I remarked that my knowledge of geology was limited to what I had learned from an old book called Aspects of the Earth. To which Gwynne replied, "If you read Shaler, I'm surprised you didn't become a geologist."

Memory of Gwynne's remark is probably responsible for my going to the library. To my delight, I found Shaler on the ISU shelves, still in good condition. The opening pages revealed that the book was already old. Published in 1910, it contained material that appeared in Scribner's Magazine in the 1890s. Shaler was a geologist at Harvard.

After reviewing the dried apple analogy, which was supposed to show that the earth wrinkled as it cooled, I detoured into the theory of volcanoes. They are dangerous. That led to the memorable letter written by the young Pliny recounting the death of his uncle in the great eruption of Vesuvius. Knowing that Uncle Pliny had a reputation for some wild tales about animals, I checked the encyclopedia. Sure enough, he said bees come from larvae that other bees gather from flowers. New colonies, however, come from a dead ox. Pliny's Natural History is described as a "storehouse of ancient errors."

Exercising restraint, I avoided turning from Pliny to plumbing or Plutarch. Back with Shaler, I landed in a section on caves and the animals therein. He appears to have accepted the basic tenets of Darwin. For the peculiar features of cave animals, however, he discarded natural selection. It was obvious that cavefish and salamanders are blind because eyes cannot develop in darkness.

That is when I turned to The Life of a Cave (McGraw Hill, 1966). Mohr and Poulson are firm with the explanation that I have generally found about the loss of pigment and eyesight in cave species. Mutations leading to these features occur both in and out of caves. Above ground, the mutant animals are handicapped. They seldom survive to reproductive age. In a cave, however, pigment and eyesight have no function. Mutants have a chance to pass their traits on to future generations. And so, we have salamanders that are blind and ghostly pale.

The account of an experiment on salamanders sent me to the library to look for another old friend, The Nature of the Beast. It is an account of Kingsley Noble's experiments in animal behavior, compiled by Mrs. Noble after her husband's death in 1929.

Fortunately, the book was checked out, sparing me additional detours. Actually, there was enough in Mohr and Poulson. They described the unique Ozark blind salamander, which has functional eyes during its larval stage. By completion of metamorphosis, the eyes have regressed and sealed over.

Noble assumed that the eyes would not deteriorate if the animals were kept above ground, and he started an experiment to test this hypothesis at the time of his death. More recently, as told in The Life of a Cave, Leon Stone of the Yale Medical School tested Noble's idea. He found that light did not save the salamander's sight. The eyes regressed on schedule.

Biology texts have very little information about adaptations for life in caves. The one exception on my shelves is Berrill's Biology in Action (Dodd, Mead, 1966). In Berril, I found that there is another angle. He tells of experiments with an African blind fish. This cave species, he says, is blind because of its low metabolism. This physiological adaptation is an energy saver that enables the fish to live in food-scarce habitat. Loss of vision is a by-product of a menu that would mean starvation for most animals.

This is a different twist. The more common view is that blind animals are not really adapted for the cave environment. They are "defectives" that are protected by darkness. If it is generally true that blindness of cave species is the result of low metabolism, these animals are really adapted. Their slow growth and development is an advantage.

Having read about the African fish, I quickly closed the book and shelved Doctor Berrill. Leaving him exposed, I might start flipping pages and come to something like floating continents or belching volcanoes.

STRIPER

When Dave F. asked if he could use the leaky 20-gallon aquarium to house a couple of ground squirrels, he wasn't sure why he wanted the rodents. "I'll just watch them and see what happens. Maybe I'll think of something."

Not a bad start. I asked if he thought they would hibernate here in the lab. He grinned and said, "Maybe that's what I wanted to find out."

After his captives gnawed through the pressed-wood lid, Dave patched it and recovered the squirrels. They burrowed down along the glass wall and carried grass into their nest. But within a few days they were out again. This happened several times. I suspect Dave enjoyed stalking his pets with an insect net too much to replace the gnawable lid by wire or glass.

Then one day we discovered that the squirrels were not only getting out of the tank. They were getting back in!

Occasionally we saw them exploring around the perimeter of the lab, skidding into a dark corner when a student came too close. To provide easier access in and out of the tank, Dave installed a wooden ramp. That's when the rodents began to spend more time on the window sill, sampling the potted plants.

Within a few days, we discovered that only one of the squirrels was returning to its nest. Inquiring around the building, Dave learned that the missing pet had explored too boldly. It had blundered into the hallway and was evicted by the custodian.

The remaining rodent established a daily routine, appearing on the work counter at the back of the room in mid-morning. It had a favorite perch from which it supervised the lab work, on a set of Seton's Lives of the Game Animals.

The thirteen stripes would blend better with prairie grasses than with the environment of room 111. And as we watched the animal slither down its narrow tunnel, the slender body, small head, and tight-fitting ears seemed just right for eluding a badger.

In its confined quarters, Dave's pet had too little space for a normal burrow. Out on the lawn bordering the prairie there are generally several openings for entering or leaving the burrow. Some of them are so well concealed that the squirrel can drop out of sight with amazing speed. I've stalked a ground squirrel with my camera, expecting the rodent to run for the only hole in sight. Suddenly it disappears as if by magic.

One reason these plunge-holes are easily hidden is that much of the digging is done from below. And even at the main entrance the digger often loads its cheek pouches with soil and carries it a distance from the opening. I have sometimes found the main entrance plugged with vegetation. Perhaps the animal does this from the outside, then enters by way of a plunge-hole.

Dave comes in every morning before classes to check on the latest developments. I suspect he is happy that his little striper hasn't hibernated. "Just watching" is its own reward.

The Wall

With the hallway connecting the west ends of the three wings, I am walled off from a view of the prairie and the valley of Squaw Creek. This is a disappointment. Yet compromises often bear some good.

The enclosed courtyard is now a semi-natural area for cottontails, woodchucks and such.

Just before the final windows of the hall were installed, we realized this was our last chance for moving large objects into the enclosure. A big log from a fallen cottonwood seemed just right for adding a bit of wildness. We lashed seven poles across the log, and 14 boys carried it, pallbearer fashion, up the hill. Easing it through the window space, they deposited it at an angle in the southwest corner. A few small boulders followed, then sumac, transplanted from the edge of the prairie.

When our rights to the space appeared to be accepted, I asked permission to dig a pond. Students brought their own spades. The school plumber installed a copper pipe through the wall of the greenhouse, and we added an underground hose. When the pond filled up, with water spilling into the manhole of the storm sewer, a trickle from the hose was enough to replace seepage and evaporation.

Cattails, turtles, and frogs; crayfish, whirligig beetles and water striders. Along with woodchucks, this is our view from the windows. Because of a wall!

Come to Order!

When science teachers are scowling at administrators - for example, when an impromptu assembly interrupts classes - Sturtevant tells us that we are not sufficiently humble. We should acknowledge the wisdom of our superiors.

Tomorrow I shall be sufficiently humble to apologize to the director of guidance, Bill Ripp. His office is in the administrative wing. This involves an incident during today's staff meeting. Before our business was complete, Adams had to leave for urgent business at a higher level, and he asked me to chair the remainder of the meeting. We were electing members for several vital committees that are assigned to consider problems such as why students are not getting the vital messages that are issued in the daily bulletins.

I rejoiced on two counts: In such a prestigious spot, I might be over-looked as a candidate. Also, I was qualified. As an undergraduate at Iowa State, I had completed J. A. Starrak's course on the nature of teaching. He was convinced that every teacher should be proficient with parliamentary procedure. We operated as an organization, chaired in turn by members of the class. I believe we considered the Seven Cardinal Principles of Education. One that I remember was worthy use of leisure time. There must have been six more.

I passed the course. But now, 35 years later, I have goofed.

My apology won't be strenuous: our director of guidance is a gentle man. The thought of it, however, reminds me of another incident in Starrak's class. He suggested that our training would be more rewarding if occasionally we confronted the chair with obstacles - an interruption in the sequence or an out-of-order motion.

Bill Stitt and I considered the options and decided on Edgar as a target . We weren't really sure that Edgar was pompous; but he had leanings in that direction. He carried a fat brief case, and his shirt pocket was well- fitted with pencils and pens. We assumed that some day he would sit in a large over-stuffed chair, his feet elevated; he would smoke long cigars. How would Edgar behave if immediately after he had called the meeting to order, we moved to adjourn?

Moved by Stitt. Seconded by Trump.

First Edgar glared at William Stitt, next at Richard Trump. Then he reddened and turned toward J. A. Starrak, who remarked, "It's your meeting." Looking pleadingly at the class, Edgar said, "It has been moved and seconded that we adjourn. Is there any discussion?"

There was no discussion. Edgar searched for help again, saying, "Isn't there any debate?" To which J. A. Starrak commented, "Motions to adjourn are not debatable."

"Then," said Edgar, "we will vote on the motion to adjourn. Those in favor say aye." There were approximately two ayes. "Those opposed say no or nay." Silence.

Edgar turned again to J. A. Starrak, who responded, "It's your meeting!"

"Then," said Edgar, "I guess we are adjourned."

We left quietly.

All of which reminds me that many years later, when we were visiting at an ISU reunion, I reminded Bill of our abbreviated class session and asked if he had made good use of parliamentary procedure. His wife Marjorie responded quickly, "I can answer that question: Yes!" And she reviewed a situation at a political party convention in Skagit County, Washington. A plank for the party platform was being debated. It stated that all 18-year-olds must register for a military draft. She said there was a lot of flag waving and it appeared that the resolution would pass. But Bill intervened with a motion to amend, changing age 18 to age 40.

Bill's amendment lost by a narrow margin. But so did the original plank resolution. "You might say both groups lost," Bill added. "But I've always thought our group won."

Oh yes - my apology! I'd better record the details as a reminder that I was once humble: Nominations were in order for a committee post. Someone nominated Edna. There was a pause, and Phil called out, "I move nominations cease." There was a quick second, followed by Bill Ripp's voice-"I have another nomination." I replied that we had a motion that required a vote. At this stage of the meeting, the mood did not favor discussion, and I called for the vote. The chorus of approval was loud, followed by a single negative voice. I congratulated Edna on her victory.

Later, on our way out of the meeting, Bill Ripp edged toward me and said, "Dick, a motion to close nominations is not in order as long as anyone wants to nominate."

Still gloating over not having another assignment, I merely nodded. But, now that I have thought about it, Bill is of course right. And one should not require parliamentary training to know that all members must have reasonable opportunity to nominate.

My apology to you too, Dr. Starrak!

PUMPKIN AWARD

At lunch today, we were discussing nominees for the annual Pumpkin Award. Although this award is not listed in the Ames High Handbook, we enjoy pretending.

The top rating this year is shared by Chuck and Al. Early in the year, they were apprehended placing heavy chains and padlock on the fire doors at both ends of the science wing. They said they were sorry.

Runner-up was Jennifer. She pilfered purses in the girls' locker room. She promised to repay.

Charles received consideration for walking calmly into the library and pressing a cream pie into the face of an attendant. This occurred on the final day of classes, and we are uncertain about the nature of his apology.

The idea for the Pumpkin Award came up several years ago on November 1. Two boys had smashed a number of Halloween pumpkins on the parking area and sidewalk .As a result of their bragging about the accomplishment, they were easily identified. Following the usual conferences, the boys were dismissed from the study hall to clean up the mess.

During homeroom announcements, the pumpkin incident was reviewed, deploring all manner of vandalism, but congratulating the boys for the clean-up. It was in the tradition of Ames High School that the boys made retribution.

That was the inspiration for the Pumpkin Award. It goes to a guilty student who, in the tradition of Ames High School, demonstrates concern for the welfare and reputation of the school.

I have a few suggestions for students who aspire for the Pumpkin Award. Don't wait until you are a senior. Start young, building a solid reputation for mischief. For a start, you might dump a half pound of garlic sauce in the staff coffee urn at the back of the cafeteria. When you are satisfied that there have been sufficient tasters, wheel in a fresh urn and a plate of doughnuts.

For your junior year, consider clipping the wires to the sound system speakers in selected classrooms. Be sure that you are apprehended before someone in the electronics class claims credit.

You are now ready for something more spectacular. Ripping off a master key to the building and selling copies to the sophomores is not recommended. Making refunds to the buyers would be a nuisance. Instead, obtain the stuffed head of a bull moose. Anchor it in the courtyard pond so the nose, eyes, and antlers are visible. For best effect, hang a strand of algae from its lips, and install a small aquarium pump so water will drip from the chin. In your apology, there is no need to admit that you received help from your old biology teacher. But be sure to promise that you will return the head to the proper moose.

UNKIND WORDS FOR BEGONIAS

Ethel has been talking to her begonias. Although the experiment didn't come out right, Ethel has an explanation.

In late November, we had completed a unit on plant growth. Too much of it was bookwork. A shortage of window space limited our contacts with living material. To provide a better-balanced diet, I asked that each student perform an experiment with living plants. This was their main homework until they returned after Christmas vacation.

"Ask yourself a question that you can answer from your own observation," I said. "Pick a question that interests you. Answer it by experiment."

Ellen already knew the answer. She needed only to show that she was right. Houseplants are sensitive to the feelings of their keeper. They grow better when you speak kind words to them.

To prove her point, Ellen started two groups of begonia cuttings in water. She said kind words to one group, hateful words to the other. Both groups flourished, and to Ellen's credit, she admitted it.

"The reason the experiment failed," she wrote, "was that I wasn't sincere. I didn't really mean those bad things I said, and the plants knew it."

OUR JUMPING FROG

"We only tested the big frog. Here are our findings. It jumped farther when it was warm."

That is how Jim B. and Dick J. reported. They were testing the idea that enzyme activity is affected by temperature. Lacking both the inclination and the facilities for working with isolated enzymes, they were willing to try my suggestion: Muscle action involves chemical reactions that are aided by enzymes. Frogs have very little control over their body temperature. If temperature really affects the functioning of enzymes, we should be able to see effect in the jumping of a frog.

So, the boys tested the big one. When pressed for details, they showed me their data. The outdoor temperature was 8 degrees C. On its first jump, the frog cleared 53 cm. In a succession of ten jumps, the distance gradually decreased to 20 cm. The average was 27 cm.

Could the decreased distances be due to fatigue? Dick didn't think so. "Frogs don't get tired that easy."

Based on the Beers-Jacobs study, we expanded the demonstration. With a little persuasion, the students agreed that they should witness the jumping, and perhaps help with the measurements. We left the frog outside in a box; its metabolism would then be adjusted to the temperature. At the beginning of each class period on Tuesday, and again on Thursday, we measured a series of 20 jumps. Averaging each series, we plotted the figures on a chalkboard graph, distance against temperature. The line of the graph climbed from 42 cm (at 7 degrees C) to 104 cm (at 25 degrees C). But there were two irregularities. The average of the jumps at 66 C was greater than the average at 7 degrees C, 51 cm, and 42 cm. Also, the average was greater at 21 degrees C than at 22 degrees C-88 cm and 75 cm.

"How do we explain these bumps in the graph?" I asked.

"It's the frog's fault," Jim offered. "It didn't always do the right thing for us either. I think it does what it wants to do."

That helped. It was another way of saying that temperature isn't the only factor that affects how far a frog jumps.

While the students were watching and measuring, I heard a number of astute suggestions: "I could make it jump farther by tickling it with an electric probe." "We should have a whole bunch of frogs and get averages." "If Hanson would stop trying those long passes and give the ball to Wilson, we'd get some yardage!"

The Stinkers

"I know these tests are supposed to make you think. The first time I took one, all I could think was I hated them. I thought they should be called 'stinkers.' Now I realize they are helping me organize what I know."

"Many people don't think in a scientific chain of thought. Questions like this pull them down."

"It's good to have questions that don't require sheer memory." "Some answers are just opinion, and when you score them you give credit for some opinions and not others. Why should we all think the same?"

"At the beginning of the year they seemed impossible. When you explained how to think about thinking, I got mad at myself for being stupid. Now they are getting easier. Maybe I am learning how to think."

"How come you use us for ginny pigs?"

Those are sophomore comments about the stinkers. And here is an example:

An Italian biologist was interested in knowing whether insects are really protected by resemblance to their environment. To obtain data on this question, he worked with the praying mantid, which occurs in two color phases-brown and green. That is, certain mantids are brown throughout life; others are green throughout life. He captured 29 green specimens and 45 brown ones. Without harming them, he scattered the insects through a patch of green grass, tying them down so they would not escape. With this arrangement, he hoped to learn the effects of protective coloration. After a period of 19 days, he found that 35 of the brown insects had been eaten by birds. All of the green ones had survived.

With regard to the praying mantids, what additional observations would you suggest in order to provide further data? Remember that the biologist was trying to determine whether protective resemblance was really a factor in the survival of the praying mantids.

That was the first such quiz that I tried, ten years ago. The results were encouraging. Less than half of the students accepted the data as proving that protective resemblance really protects. And even among those who accepted the evidence, some recognized that the experiment was incomplete. They reasoned that green mantids may taste bad, or that the green ones may fight back and escape being eaten. In brief, mantids may differ in ways other than color. Good! The experiment needs a control. And many students suggested exactly what the experimentor, de Cesnoli, actually did. He tethered another batch of mantids on brown grass. In this part of the test, the green mantids were eaten, and the brown ones survived. There was a variety of other suggestions. One was to place mantids of both colors in a situation where neither color resembled the background. Such answers were accepted.

A more difficult question concerned the geographic distribution of organisms:

Many kinds of native American plants have scarlet-red flowers. No such plants are native to Europe. This aspect of plant distribution has been explained by the distribution of animal pollinators, as follows:

The honeybee, which is "color blind" to red, is a native of Europe but not of America. The hummingbird, on the other hand, is a native of America but not of Europe. Some kinds of plants depend on animals for pollination of their flowers. The hypothesis is that the plants and the pollinating animals have gradually become more specialized and more dependent on each other.

The distribution and characteristics of these plants are therefore closely related to the distribution and characteristics of the pollinating animals.

Concerning the distribution of plants with scarlet-red flowers, what additional evidence would make the above explanation more convincing? Limit your answer to evidence that could be obtained by observation or experiment.

The missing evidence here is the color sensitivity of the hummingbird. If the bird, like the bee, is color blind to scarlet red, the theory about the distribution of the red flowers is weak. If the bird is sensitive to this color, the theory is much stronger.

Of course, there were complaints. "How were we supposed to know what that Italian did with his bugs?" "We never studied the eyes of hummingbirds!" I heard from parents, too. I should not expect children to know something they hadn't been taught.

Although the problems are new, the principles should be familiar, and they are related to the topics being studied. Protective resemblance comes up in early autumn, when we are observing adaptive features of grasshoppers. I used the red flower problem after the students had learned about the limited distribution of organisms over the earth. And here is one related to sensory functions:

Catnip, a plant of the mint family, produces a substance called nepetalactone. This is the substance that causes such odd behavior in cats. They eat the leaves excitedly, and rub their bodies against the plant. Biologists have puzzled about the value of nepetalactone to the plant. As one biologist observed, "Surely a mint plant gains nothing from being able to excite cats."

One hypothesis suggested that nepetalactone might protect the plant from insects. Attempting to test this idea, he placed some of the substance in a glass tube. When he held the tube close to various insects, they moved away. In fact, he found that he could chase the insects around in any direction with the tube of nepetalactone. How do you think this test of the hypothesis might be improved? Write a brief explanation.

The clearest control here is to find whether the same insects retreat from a tube that does not contain the catnip material. An encouraging number of my "ginny pigs" seem excited by this kind of challenge. Is it really a new experience, using biological information instead of merely repeating what the teacher has said? The most startling revelation to me, however, is that many of my sophomores believe that the type of thinking required in these tests is useful only in the study of science.

Ruth Ostle, who did her student teaching in my classes, tells me that I resorted to trickery to keep some of my students occupied while I graded notebooks. She says I assigned small groups to monitor the number of times our Squaw Creek minnows blinked their eyes during five-minute intervals. Is it possible that I was bearing down on the absence of eyelids in fishes?

Whatever my motives, how could I forget a technique that must have worked so convincingly?

Bodiless Feet

During our Christmas program, the auditorium lights were dimmed and the stage darkened for a change of scenes. This is a critical situation for the behavior of a high school audience. Loud voices and paper airplanes from the balcony are among the kinds of disorder that may erupt. In this solemn event, however, order was holding. Then Bob S., a stagehand, walked out on the darkened stage wearing a pair of glow-in-the-dark shoes. Seeing those bodiless feet moving around, the students broke up. Herb then came out and explained that this was a sacred observance. Silence followed. But, as Bob walked off, laughter again broke loose, and the remainder of the program was cancelled.

Those Thirsty Roots

No one responded. I repeated the question: "Now that you have seen what happened, were the roots searching for water?"

Elaine A. looked as though she was about to speak; she raised her hand in a cautious gesture, then withdrew the offer. I looked for another. Finally, Mike B. grinned, "Well, if those roots were really looking for water, I don't think you'd still be trying to get us talking about it."

That would have been a good time to say something clever. But I laughed with the class and asked Mike if he could show us anything in the growth of the roots to back up his answer.

"What I see," he replied, "looks like the roots are trying to find water. That's why I think you want us to see something else."

Elaine's hand was edging up again. I avoided seeing it and asked the students to inspect the growth chamber again. The glass walls of the chamber slant inward, making a V-shaped cross-section. Because roots generally grow downward, this design causes more of them to press against the glass walls. We had inserted four metal pipes, two at each end of the chamber, close to the glass. At one end, the pipes were perforated by small holes, so water could seep out. At the other end, the pipes were fully water-proofed. We started the experiment with uniformly moist soil. On one side of the chamber, near the wall, we planted corn seeds; on the other side, pea seeds. By the time the seedlings were breaking through, we used a funnel and added water, but only through the four pipes. At one end, the soil became dry. At the other end, the soil was moistened by leakage from the perforated pipes.

In both kinds of plants, the branch roots near the perforated pipes grew faster than others. The roots certainly appeared to be growing toward the water source.

It isn't easy to be objective, explaining such observations. This is especially true because it is common knowledge that drainpipes and sewer lines are often plugged by roots that are "looking for water."

After the students had examined the corn roots again, I turned the growth chamber so the peas were visible. Then I looked at Elaine and asked if she was ready to talk. "The plants don't look so good near the closed pipe," she said. "Maybe it's because the soil is too dry. At the wet end, the plants look better, and the roots are longer. Maybe they grow near the pipe because the water is there, not because they are searching for it."

Thank you, Elaine. And thank you, Mike. You are both right.

One of my clippings from the Des Moines Register illustrates how the teleological view of life can distort the facts. There had been an extremely dry summer in western Iowa. In a story on soybean yields, the writer quoted agronomist Howard Taylor as saying, "Roots on those soybean plants had gone down eight feet in depth searching for water. We had known that some soybean varieties would go to depths of five feet, but this was the deepest we had seen because it was such a very dry year."

Searching for water? I called Dr. Taylor. He said that he had been interviewed by telephone, and misquoted. His explanation: early in the growing season, there had been enough moisture to give the roots a good start. Then came the drought. The upper soil level became so dry that development of the branch roots was seriously retarded. But there was still moisture available at deeper levels where the roots had already penetrated. More of the food manufactured by the leaves moved down into the roots where the water was still available. The roots grew where water was, not where they hoped to find it.

For many students, it is a subtle distinction, growing better in the presence of water as opposed to searching for water. Sometimes I wonder why I strain to get this point across. But as is true of many teleological statements, the search idea obscures the real explanation. It encourages thinking of a root groping along a water pipe, looking for a leak.

Bad Therapy

There are two convenient explanations for the failure of students to grow and prosper as students are expected to grow and prosper. One is faulty instruction. Another is a defective textbook. Most teachers choose the latter. Blaming the book is good therapy.

Now that I am using Design for Life in my honors class, I am in trouble. Both choices are bad. Today, one of my bright students put my problem in bold print. With back-to-school night coming up, I asked for suggestions in talking with parents: "I have only 15 minutes to explain everything. If you have a suggestion, leave an anonymous note on my desk."

One of the notes said, "Just tell them why we can't understand the book you wrote."

BRAIN PICKING

At one time, it was called use of community resources. When there was a "felt need" for improving the language of education, it became utilization of available talent. More recently, it merged with other phases of progress as innovative thrust. I am still satisfied with "picking local brains."

During my first year on the Keokuk staff, with my own training heavily agricultural, I leaned on local physicians. It started even before my first day of classes, when I told Dr. Holmes about my assignment. Along with several prescriptions for treating quirks of the human mind, he presented me with a stethoscope. From that day, I knew that it pays to ask for help.

I was reminded of this today while sorting through some old reel-to-reel tape recordings. As a refresher, I listened to snatches. In a recorded interview, Dr. William Baird told a group of students that bones are bloody organs. How else could they grow? How else could they heal injuries or produce red corpuscles? He was answering a question about the little holes near the ends of a human femur that the students were examining.

Dealing with smoking and heart problems, Dr. George Montgomery explained that most heart attacks are the result of a poor blood supply to the heart muscle. Nicotine, he said, causes blood vessels to constrict. "If your blood vessels are already doing a poor job for the heart, nicotine reduces the flow still further. If part of your heart muscle is damaged by lack of oxygen, you are in real trouble!" This was years before the Surgeon General put his stamp of disapproval on cigarettes.

Extension Specialist in pest control, Harold Gunderson, warned us that Ames, too, has a rat problem. He said every town has a rat problem. If we want to poison rats, we must pre-bait, feeding the same material in which we will later add poison. Rats are wary animals. They sample new foods cautiously. The small amount of poison they would get in their sampling would only make them avoid the food.

Lugging a 60-pound recorder to an interview was a bother, but having a permanent file of the conversations was worth the trouble. And when small cassette recorders came along, we were really in business. I found that busy professionals could not be expected to take a day off to visit Ames High. They were generally happy, however, to visit with one class or with a committee of interviewers for an hour.

Many consultants prefer to have a list of questions in advance. It is easy to comply. I remind classes of what they have been expected to learn in connection with the interview topic. Then I give them a chance to ask questions to be answered later by a specialist. Having the questions in writing, I can edit and arrange them in logical sequence.

There is generally no problem finding a small group of students to conduct the interview. They feel secure, knowing some of the questions in advance. But I urge them to ask the questions, not to read them. "Make it appear that you just thought of your question, and feel free to inject comments or further questions as the conversation moves along." Make it a live program."

Many of the interviews, I've found, are more useful in classes if transcribed onto paper. This permits editing, getting ideas in logical sequence. Also, it permits the consultant to approve the transcript before it is duplicated. On reading his or her own words, the consultant occasionally finds that further explanation would be helpful. To my surprise, many students would rather study an interview on paper than on tape. Perhaps they feel more in control, reading at their own pace and taking notes without losing the thought that follows.

Listening to old tapes, I was amused to discover that I had forgotten the source of some of the information. I had used it enough that it had become a part of me. Perhaps that is one of the good things about picking people's brains.

Note in my mail box: "Call Mrs. Graham, 232-4573. Raccoons in chimney. Will be home most of morning."

On my way down the hall, a little late for class, I met Mike R. He was headed in the wrong direction, holding his hand over his face. "Nosebleed?" I asked. "No," he replied, "but may I go to the restroom? I was smelling Kim's lipstick, and she pushed it up my nose." Request granted.

Comment from a committee member: "If we are going to get anything out of it, we are going to have to put in more input into the planning." I am unsure that this will really affect what comes out of the output.

UMATLD

I am now prepared to announce a superior control for urban mosquitoes. It may be considered a fourth generation approach. (The first was to smash the bloody insects; the second to drain the drainage ditches; the third to sprinkle lethal emulsions on the waters.)

My proposal, known as Urban Mosquito Abatement Through Larva Destruction, involves providing abundant breeding habitats. Every civic-minded household will establish one breeding site for each person. Each site will require a waterproof receptacle, holding upward of a pint. Such items as pickle jars, coffee cans, plastic whipped cream bowls, and wading pools will qualify.

INSTRUCTIONS: 1) Distribute the breeding containers over the property and add water at 7 a.m. Monday. 2) In mid-week, say at 8:20 p.m. Thursday, examine the breeding sites. If anything wiggles, take the container to the rose garden and pour out the water. 3) Replace the container and add water. 4) Repeat the procedure at 7 a.m. on the following Monday.

DISCUSSION: Although most mosquitoes require at least a week to complete the life cycle, some species do so in a few days. It is therefore crucial to water the roses regularly. Urban Mosquito Abatement Through Larva Destruction, known also as UMATLD (pronounced umatld), has the advantage of adding favorable organic matter to rose gardens, using inexpensive equipment, and avoiding pesticides that kill dragonflies. In addition, it interrupts the life of a female mosquito before she is able to deposit her eggs in neglected pickle jars, coffee cans, plastic bowls, and wading pools.

Friday

My defective ear is a superb advantage at times. Sophomores believe I don't hear what they're saying.

Spokes

With a fresh degree in music education from Drake, Patty has been substituting for ailing teachers in the Des Moines schools. By way of passing on the old advice about starting with stern discipline policies and easing up gradually, I said, "Don't smile until Christmas." Her reply, "Substitute teachers never smile."

We were talking about reading problems, and Floyd kept asking if we had any suggestions for the committee. Finally, I said, "Tell them the first step is to open the book." To which Jim added, "Open the book and shut the mouth!"

Charlie Windsor was recounting one of his early tangles with education. As he understood the teacher's order, he was to go to the back of the room and stand in the wastebasket. It was a tall wastebasket. Charles was a short student. After valiant efforts to get into the container without upsetting it, he looked to the front of the room, hoping the teacher would come and help him in. But the teacher was sitting there watching, and laughing. Finally she walked to the back of the room and whispered, "Charles, I told you to stand by the waste basket, not in it!"

Some days, while riding my bike to school, I watch the cars that are passing me, and I think that it would be a lot faster to drive. Other days, I glance down at the front wheel spokes and think that this is a lot faster than walking. Kids undoubtedly respond to the mood of the teacher. Perhaps I could predict the quality of my day by noting my thoughts en-route to school. Am I watching the cars go by or seeing the whirl of my own spokes?

Curious Kids

Every time an elementary class visits my lab to watch the bees or to examine Mr. Trump's skeleton, I wonder what happens to the curiosity of children as they grow up. It goes somewhere. First graders - even fifth graders, to a degree-are so full of wonder, so full of questions. Would kids remain curious if they were not taught? I do not know.

By taking one of my sophomore lessons into three elementary classrooms, however, I observed the persistence and penetration of children's minds. At Ames High we had been using a growth chamber for the study of roots. It permits seeing the response of roots that are confronted by a rock or by the hole of a burrowing animal. It provides a vivid demonstration of the difference between the fibrous roots of grasses and the taproots of carrots. Such facets of life are generally hidden beneath our feet.

Having seen how quickly sophomores can satisfy their curiosity, I arranged with three elementary teachers to set up a demonstration in their classrooms. The growth chamber was stocked with corn seeds that were germinating. Alternate seeds, four inches apart, were placed with the root end down, the others with the embryonic root pointing up. On each side of the chamber, I had glued a flat stone to the inner surface of the acrylic wall, directly in the path of a developing root.

As agreed, the teachers did not comment about the sudden appearance of the demonstration. If asked about it, she could generally get by with a reply such as, "Perhaps you can answer your own question just by watching. Then next week you can tell us what you decided."

For two second grades and two third grades, I visited the class ten days later. The teacher talked informally with the children, asking what they had seen and learned. I taped the discussion. After about ten minutes, the teacher introduced me and suggested to the children that they might like to ask questions.

Many of their questions were more like answers: "We had some corn in our garden, and it got bigger than my grandma!" "My mom thinks I don't chew my food enough." "My daddy has a whole bunch of corn seeds, but he gives them to the pigs."

There were a few real queries: "Where does the seed go when the corn grows up?"..."How does a root know what to do when it is upside-down?" And one alert little fellow asked where seeds come from. I replied that all seeds come from flowers, and on a corn plant the flowers are hidden by the husks. The boy wasn't satisfied: "I mean where did the first seeds come from, before there were any corn plants?" A little girl broke in with a comment about her mom making dolls out of cornhusks. The interruptions helped me handle a question that seemed better suited for a biology class than a second grade.

I learned more, however, from the teachers. Barbara Grossman, second grade, said her pupils reacted "as positively to this experiment as to other science activities." They didn't talk much about the plants. But they inspected the display frequently and were excited when they discovered an obvious change.

What changes? There were spurts of growth in both roots and stems. This was apparent because on alternate days the growth chamber, on a projection stand, was wheeled into another second grade classroom. When roots turned almost at right angles, growing around obstacles, the children were delighted. Some of them failed to notice that when a seed was placed upside-down, both stem and root made an abrupt U-turn. Others reported what happened: "When it was about this long, it just turned and went the other way."

The other second grade teacher, Sherri Bohning, was pleased by the quality of observations in an undirected activity. The children could enjoy openly, without guidelines for success or failure. "They were watching a miracle," Sherri continued. "They could sit there, observing pensively and wondering. They could interact without competition. They needed only reassurance that their ideas were all right."

One of the surprises to me was that children of this age need opportunities for seeing the relationship between time and change. Most of their observations of living things involve quick changes-the fright reactions of a baby, the cuddling responses of a puppy. "The growth of plants," said Sherri, "gave the children a feeling of time that they don't normally encounter." Their principal Dale Brentnall agreed. He said adults seldom recognize that children must learn to observe slow changes.

For contact with third graders, I took the growth chamber to Roosevelt School and followed the same procedure as at Fellows Elementary. The additional year of mental development was evident in the continuity of the questions and observations: "I learned that seeds won't grow after they are cooked." "The way to get corn off the cob is with a combine; it takes the seeds off and throws the cobs away." "A tree root was so tough when a car tried to pull it out, the chain broke!" And "tiny seeds can make big plants if there is water and sun."

Again, the teachers' comments helped. Nancy Beavers explained the advantage of having the demonstration available for an extended period. Short-term lessons, she said, don't allow time for the children to reinforce their observations. She believes having the plants in view on alternate days for a ten-day period was about right. Many of the children saw a relationship between their observations and other experiences.

Sharon Ortigies was impressed with the value of incidental learning in the third grade. "We can't expect most children of this age to depend on books. Their path to science is in first-hand experience." Claire Allen's fifth grade at Crawford School had the demonstration to themselves for several weeks. I did not visit the class, as I had done for the lower grades. Claire said there were not many surprises for her pupils. "They seemed to know what would happen. But they checked the progress regularly. And they were immensely pleased to find that the roots were doing what roots are supposed to do. This reinforcement gives children confidence in themselves."

Claire reminded me that her pupils were ready to go beyond casual observations. They plan experiments to test their ideas. I talked with Roosevelt's principal, Tom Berhow, about this, asking at what age children think in terms of experimenting. He said third and fourth graders could certainly use the growth chamber for simple comparisons. And they can appreciate quantitative results that are presented in graphic form.

Faculty Hand

Farrar told us he'd be less than honest if he failed to say that student conduct relative to the hallways had deteriorated. He said we should either enforce the regulations or change them. He waved a black notebook and said, "This is the faculty handbook!"

To which Hank Hanson quickly responded, waving his fist in the air. "And this," he said, "is the faculty hand!"

Step Test

The rhythmic commotion in Spatcher's lab turned out to be a test of physical fitness. Half the students were stepping up on their chairs in unison while Spatcher called out up down up down. The remainder of the students were holding the chairs steady.

Sensing my curiosity, Spatch pointed to his kinesiology textbook, opened to a section labeled Harvard Step Test. I tried it in my classes. It was good.

The Harvard Fatigue Laboratory was commissioned by the Armed Forces to standardize a test that could be administered easily to large groups of inductees. It lasts only five minutes, for those who can "take it." The score is based not only on how long you can endure, but also on how rapidly your heart rate recovers from the stress.

Before using the test next year, I will record the verbal part on tape, supplementing the voice with the tick of a metronome. Then I'll be free to supervise. Also, I will caution students not to participate if they've been ill or have been advised to avoid strenuous activity. It is really strenuous!

The Great Grasshopper Hatch

The last of our caged grasshoppers died about the time the pond iced over. There was plenty of food. But even when protected from October frost and November ice, their life spans had run out.

Now it is May, and we are again tabulating data showing that life is often more complex than it seems.

The project began four years ago when we were talking about adaptations that permit organisms to survive Iowa's winters. Adult grasshoppers die with the advance of winter. This much the students had learned in the third grade, from the story of the grasshopper and the ant. As sophomores, they learned that the hoppers survive our winters as embryos, in eggs. And from one of our books they "knew" that in this climate, hopper eggs won't hatch until they've gone through a period of freezing. A neat adaptation. It prevents hatching in autumn, leaving the young insects to face a frozen future.

So we decided to demonstrate this adaptation. We collected mature specimens of Melanoplus differentialis, the big yellowish ones, and confined them in two narrow glass-walled cages. We added grass and other leaves each day, and occasionally sprayed "dew" into the cage as a source of water. We watched them copulate, the smaller male twisting his abdomen around to insert the sperm. And sometimes we saw the female dig her ovipositor deep into the sandy soil. Once a female dug in against the glass wall, and we could see the abdominal contractions as the eggs piled up and were covered with a frothy liquid that hardened into a protective pod.

When the last of the hoppers died, we sifted the egg pods from the soil and divided them into four groups of equal mass. Covering the pods at normal depth in four jars, we labeled one test and left it in the lab. The three controls we placed in the refrigerator for a few days, then in the freezing compartment.

We moved one of the jars back into the refrigerator for gradual thawing on January 20; then brought it into the lab. Another came out of the freezer on February 28, and a third on April 6.

The results? Apparently, M. differentialis doesn't care what the books say about temperate climate grasshoppers. The average number of hoppers that emerged from eggs that had been frozen was 349 per jar. The average for those not frozen was 303 per jar. There was an interesting difference, however, in the duration of the hatching period. It was longest for the test groups, averaging 132 days. Those frozen until January 20 averaged 41 days; for the February 28 group, 26 days; and for the April 6 group, only 14 days.

In ten of the jars, hatching began about four weeks after removal from the cold environment, 24 to 31 days. In two jars, the timing was quite different, 15 days and 45 days. There is a likely explanation for the slow one. When the jar was removed from the freezer, it was kept in a spot where the temperature did not rise above 62 degrees F. For the first ten days, and during the following week, the temperature was a little below normal.

Temperature wasn't the only factor that may have influenced the hatch. Grasshopper eggs are probably sensitive to variations in soil moisture. We tried to avoid both dehydration and waterlogging. But this was difficult to regulate. Also, though the egg pod totals in the four jars were of nearly equal mass for anyone year, we do not know that there were equal numbers of viable eggs. The mass of the pod coverings may have varied. Parasites may have been at work in some, not in others. Even so, we know that in this species the autumn eggs do not require freezing. The period of dormancy, diapause, is apparently of low intensity. It is easily broken, and development of the embryo resumes. Yet this restraint is enough to protect the species from the hazards that would follow autumn hatching.

Those Stinkers Again

The survey of sophomore opinions about my experimental thought-questions prompted me to find how the scores compare with those on my other tests. If there is a high correlation between the two types, they may measure the same intellectual qualities. If the correlation is very low or negative, they measure something different.

Now I know. The university's testing service put the two sets of scores for 120 students into their computer. The correlation is high, 0.813.

Therefore, if tests served only to provide grades, the stinkers would not be needed. I believe, however, that they are much needed. They help me see how my sophomores are thinking. They trigger conversations that would not otherwise occur. And they force some students to discover that it is possible to put seemingly unrelated ideas together, to come up with new insights into life.

A Lean and Hungry Look

"A Lean and Hungry Look; He Thinks Too Much; Such Men Are Dangerous" - W. Shakespeare

I knew from the start that I must watch that boy in the outside seat of the third row. He was too quiet. His face was thin and placid - too placid. It was hiding trouble.

Today my fears were confirmed. He was wearing a shirt with words on it. As he turned to sit down, I saw, "Make America look ridiculous..." There were more words, but I missed them. That boy needs help.

After the students were busy on an inventory test, I walked to the back of the room, then turned to look. Part of his shirt was hidden by the back of the chair. All I could see was, "Make America..." Pondering what sort of revolution he was planning, I checked his enrollment card. In response to my question, "Why are you taking biology?" he had replied, "Because it is on the list." Already, I thought, he has learned to be evasive. He will become one of those dangerous men, lean and hungry, who think too much.

As the period ended, I stationed myself for a clear view of his back. And now I know: "Make America look ridiculous. Ride a unicycle."

What's Good About the Male

The trouble with teaching about reproduction is that sophomores are interested in sex. They know too much. They don't listen.

I have struggled to help them recognize that reproduction has an adaptive function that goes beyond making new individuals. It is mainly responsible for spreading variations within a species, variations that permit the species to adapt to slow changes in the environment.

If average yearly temperatures drop, and ice sheets encroach, for example, organisms cannot change their heredity to fit the weather. The species must have individuals that are already genetically adapted for cold. They are adapted to survive, and their offspring pass the desirable traits on.

How do sophomores respond to the prospect of a thousand years? They are concerned with now.

The point is that many kinds of living things can reproduce without sex. A paramecium squeezes itself apart and becomes twice as many. A raspberry stem droops over, touches soil, and takes root-a new plant is on the way. There is a sea anemone that sheds its tentacles. These pieces of life drift in the water, then settle on the bottom, and each piece is capable of becoming a whole animal.

Why hasn't nature settled for this way of life? After all, it reproduces. Efficiently!

The reason is that one-parent productions are too uniform. They offer few alternative routes to survival.

With a sharing of genes from two sexes, however, a species is more adaptable. That is why two-parent productions are so common, even in species in which a single parent can go it alone. Two paramecia get together and blend their heredity by conjugating. The raspberry plant produces flowers that are cross-pollinated by honeybees. The sea anemone dumps sperms into the ocean, and some of them miraculously reach the eggs of another sea anemone. In fact, the potential for two parent reproduction is nearly universal.

This suggests that the male has something worth offering.

And there is more evidence. Certain animals as well as plants have both male and female parts in one individual. Yet there is generally a design or behavior that prevents self-fertilization. Two earthworms meet part way out of their burrows on a damp spring night. With their bodies overlapping, facing opposite directions, they trade sperm. One mating; two families. Milkweed pollen is delivered from flower to flower by flies and other insects. There would be less risk of failure if the pollen fell on the stigmas of its own flower. But, as in the earthworm, milkweed takes the route that assures a more adaptable species.

Most sophomores are amused, but not impressed, by a significant difference between the reproductive strategy of humans and of most other organisms. In present cultures, at least, the human problem is keeping the sexes apart. The sex urge has few time restrictions.

In many species, however, the problem is getting the sexes together. Males and females often live apart. If reproduction is to occur, they must be programmed to meet at the right time. And for some, it must be at the right place. Toads are an example. The males and females live solitary lives, except during the breeding season. Then they must keep their appointment at the pond, where the female releases her eggs and the male spills his sperm over them.

Among many spiders, courtship is a hazardous affair for the male. He is about the size of the insects on which the larger female preys, and she apparently is not highly skilled at counting legs. If the male's courtship signals fail, he may become a meal instead of a mate.

If I mention the intricate adaptations that insure two-parent reproduction in plants, or if I refer to the amazingly specialized courtship in animals, students suspect that I am evading the real story. They believe I am building up to rehash the perils of high school pregnancy or the hazards of syphylis. They yawn.

Good Old Yesterday

There are days when I suspect I was doing a better job teaching biology years ago, when I knew less about teaching and less about biology. It's a distressing thought.

Consider honeybees, for example.

When I started teaching, I was able to distinguish a worker from a drone. And I knew that drones don't sting. That much I had learned in my own high school days at Hamilton. My zoology class visited the Dadant Company. I took a few drones back to school and quietly released them in the study hall. My impression is that Miss Callahan did not approve of drones. Thinking they were flying in through the open windows, she sealed off their means of escape and stalked them, armed with her Latin book. I haven't shared this bit of history with my classes.

Despite my ignorance as a young teacher, students were curious and cooperative.

I studied an article in an old issue of National Geographic, and I was confident. We examined the intricate adaptations of the worker's tongue, her legs, her sting. We spread out over the neighborhood, following individual bees from flower to flower, finding that each bee was faithful to a single kind of plant. It did not waste dandelion pollen on apple or apple pollen on tulips. I found that I hardly needed a lesson plan. The student's questions provided continuity.

My answers, however, were shaky. Here are a few examples.

Question: How does the queen know that she is supposed to lay a fertilized egg in a worker cell, an unfertilized egg in a drone cell? Wrong answer: She doesn't have to know. When she lowers her abdomen into one of the large drone cells, the egg-laying tract is in a different position than for the smaller worker cell. The egg automatically gets past the opening that would allow a sperm to reach it. Right answer? Who knows? Koeniger's research shows that the queen actually measures each cell, using her front legs as calipers. When Koeniger amputated one front leg, the queen made mistakes. I think the wrong answer is easier to believe.

Question: If swarming is so bad, losing all those bees, what can you do to prevent it? Wrong answer: It's easy. You clip off part of one wing from the queen. Then she can't fly, and the swarm won't leave without her. Better: Actually, there are a lot of answers, but not enough to really solve the swarming problem. Swarming is the way honeybees reproduce colonies. It's difficult to persuade them to give up something that comes naturally. Providing proper space for brood rearing and food storage will help. But there are many factors that influence swarming, and they are difficult to control. As for wing-clipping, it's true that a swarm won't go far without the queen. In their preparation for swarming, however, the workers will waste a lot of time loafing when they should be harvesting nectar. Also, a swarm may escape with a new virgin queen that was reared as a replacement in the home hive.

Question: How do bees know what kind of stuff they should bring back to the hive? Incomplete answer: It's inherited behavior-instinct. All we know is that it works. Better: They don't always know. They make mistakes. When pollen is hard to find, the bees may collect sawdust, coal dust, granules of insecticides, many kinds of particles that are about the size of pollen grains.

Question: You said the cells of the honeycomb are used for raising young bees as well as for storing honey. If you want to harvest some honey, how do you keep from getting bee grubs in with the honey? How I answered: Fortunately, the queen stays in the lower part of the hive. Honey is stored in the upper part, and that's where you get the honey for yourself. How I should have answered: As activity in the hive steps up in early spring, you prefer to find the queen working in the lowest story. If she isn't there, you shift the position of the boxes, placing hers on the bottom. When that box becomes crowded with eggs and brood, the queen will move up. By shifting the position of the boxes, you confine her to the lower two or three boxes. (By the time a second box becomes crowded, the bees have mostly emerged from the previous one, and it can again serve as a nursery.) If you fail to make adjustments of this sort, the queen may move up and place eggs in the frames intended for your share of the honey harvest.

Another way: You can also keep the queen down by placing a queen excluder above the brood nest. It is a grid with openings large enough to let the workers through, but too small for the queen. She stays below the excluder. Careless use of an excluder, however, may lead to swarming. And some beekeepers think it slows down the traffic of the workers.

Reviewing my ignorance gets embarrassing. I can tolerate these thoughts, however, if they will help deal with the present. While I still have a few years left, I should find why it is more difficult now to stir the sophomore mind.

Could it be that poor answers are better than good answers? If so, I should still be able to motivate. I still have plenty of shaky answers.

In fact, I recall a recent explanation that shakes. Two years ago, we lost three of our four colonies of bees to winterkill. When we found the colonies dead in the spring, I told my classes that the prolonged deep snow, drifting around the hives, had restricted the oxygen supply. The bees had suffocated.

Now that I've been reading the bee journals more diligently, I suspect there were conditions more serious than oxygen depletion. There was no ice crust, just soft snow.

Although the snow may have permitted plenty of oxygen-carbon dioxide exchange by diffusion, it certainly confined the bees in their hive. Normally, bees leave the hive on mild days, relieving themselves of digestive wastes during short flights. If they are forced by long confinement to defecate inside the hive, this may lead to the spread of an intestinal parasite, Nosema, a real threat to the colony.

Along with disease, prolonged cold presents another hazard. The bee cluster has difficulty moving to new comb areas for food. Even when there is honey and pollen only a few frames away, the bees may starve.

And, finally, water vapor accumulates in the poorly ventilated hive. It condenses into drops on the hive cover, then drips down onto the combs. In very cold weather, the moisture is deposited as frost that gradually closes in on the bees.

The suffocation answer is easier. It sounds authentic, too. But science is the search for truth, not the dispensing of easy answers. That's another way of saying that my responsibility is to excite students about learning, not about knowing. This conviction has come gradually, perhaps through C.A. Lawson at Michigan State and Chalmer Roy at Iowa State. I wonder. Have I overworked this view of science?

There is of course the possibility that the students have changed. On bad days, this is a favorite theme of tired teachers. Too many of our Ames students suffer from a false sophistication. They've seen the educational displays at Iowa State's VEISHEA. They've walked through the Museum of Science in Des Moines, the American Museum of Natural History in New York. They have traveled abroad. Many of their parents are Ph.D.s. The kids are about to get a driver's license, and a car is waiting. And a disturbing number of them know about the "mind-stretching" effects of drugs. Who needs curiosity?

Even if I conclude that students are different today, I am different too. My hair is gray. Suppose this were the Monday of my career instead of the Friday. Equipped with my present knowledge, and with all of the techniques I have borrowed from successful teachers, would I be able to stimulate the sophomore mind more effectively than I do in my gray-haired, trifocal condition?

Perhaps it is better that I don't know the answer.

And perhaps tomorrow will be better.

Early November

November has trouble deciding on a season. One day it is the summer of Indians; another, the gusty chill of March. Yet I know this is autumn. The leaves of our birch are down. Our white pine has lost a crop of brown inner needles, and the tree now stands in the full green of younger growth at the ends of the twigs. Boxelder bugs are crowding the crevices of the woodpile on the terrace. The juncos are with us again, sampling the seeds of the wild lettuce.

On November 2, while replacing the wren house with the bird feeder, I saw a pair of grasshoppers in their twisted embrace. They were on the post where I was mounting the feeder, and I teased them into a can without disturbing their urgent business. They remained together when I transferred them to the egg-collecting cage. In fact, they were still entwined the next morning.

I get the impression that life cannot afford to waste an opportunity. It begets more life.

When I see how the tiny elliptical eggs are packed into one of the "pods" by a female hopper, I wonder whether this productivity may be extravagant. Two hundred eggs from one female must be a modest guess, and her quota for maintaining the species is only two grasshoppers that grow to maturity.

Then I remember that grasshoppers have enemies more dangerous than exterminators who go around sprinkling poisons on the hoppers' favorite food, more dangerous than sophomores, who go around grabbing them and putting them in peanut butter jars.

Several kinds of blister beetles make their living during their larval stage by digging into grasshopper egg pods, devouring the next generation. And there is a grasshopper bee-fly to which entomologists have given the vicious name Systoechus vulgaris. Its larvae have a similar appetite for hopper eggs. The adult female, which looks much like a honeybee, flies around in grasshopper country dropping her eggs, one at a time. Those that fall near egg pods may further reduce the grasshopper population for the following year.

Hoppers that escape the egg predators may nibble on plants that are infested with the eggs of hairworms. Emerging in the host's stomach, the young worms drill through the wall into the body cavity. There they live as parasites. Even though the insect may live until the hairworm bores its way through the body wall (to mate in the soil and deposit more eggs on leaves), the insect is weakened and infertile.

We often find these nematodes in the egg-collecting cages, and in jars that served as hopper homes on their way to school. This helps explain why we find fewer egg pods than we expect when we sift the soil in the bottom of the cages.

Years ago a boy brought one of the hairworms to school in a bottle of water. It was wiggling its way around in lively style, and the boy wondered if it is true that if you place a hair from a horse's mane in water it will turn into one of those "horsehair snakes."

In September, when we were using grasshoppers for lab observations, I asked that each student bring in a large living specimen. There were the usual excuses: "My little brother was petting it, and it jumped." "The cat ate it." "In the morning it was gone. I guess it got away or something."

Ann had a particularly sad story. "I looked and looked," she assured me. "They always jumped before I got there. Then I finally saw one sitting still on top of a weed. It was a nice big one, and I sneaked up on it. But when I grabbed, it just fell apart in my hand. I guess it was dead."

I agreed. And I guess that Amy's hopper was the victim of a lethal fungus that takes a heavy toll of its hosts during warm humid weather. Fungus strands invade the insect's innards, destroying its vital organs. With the approach of death, it climbs a plant and clings near the top by its front and middle legs. At death, its rear legs are often extended. Before the grasshopper dries, reproductive strands of the fungus grow out onto the surface. They produce tiny spores that are shot into the air. If such a spore lands on another hopper, well, that's more trouble for Amy and her friends. And that is one more reason why Melanoplus differentialis had better take every opportunity to mate and leave some viable eggs in the soil - even in November.

Are Notebooks Worth the Trouble?

Though I have heard countless sophomore opinions about the utility of notebooks, I've had no quantitative record of those opinions. To fill the void, I conducted a survey. Assuring my classes that I wanted their ideas only as a way of improving the use of their time, I asked that they sign their names on the questionnaire. This enables me to find whether the answers were different for high achievers than for low achievers. There were 40 students in the A-B range, 54 in the C range, and 32 in the D-F range.

Those amazing D and F students! In spite of their almost always careless and nearly useless notes, 87% answered "Yes" in response to the question "Do you think we should have more tests in which you are permitted to use your notes."

My Last Principal

Yesterday I wrote a brief note to principal Ralph Farrar, suggesting ways of saving energy around Ames High. Closing the windows of the history room, across the courtyard from my lab, on cold winter days, would help. So would turning off one string of lights in each hallway. (Students don't require that much light to do what they do in the hallways.) To clinch my point, I typed the message on a tiny scrap of paper, leaving only a fractional margin. In longhand, I wrote, "This paper was recycled from a principal's bulletin."

Today there was a reply in my mailbox, written on a used envelope and stapled to my note. It said, "Mr. Trump, I am returning this so you can use the other side."

The lights continue to burn; winter still ventilates the history room while the thermostat kicks the temperature higher; and quarter-page memos flow from the office on full sheets of paper.

Ralph Farrar possesses in full measure a trait that must help preserve principals. He can listen politely to opinions of teachers, then go his own way.

When Farrar was selected to succeed Herb Adams, rumors indicated that he favored modular (flexible) scheduling. To me the modular plan seemed to chop the workday into little pieces, stir the pieces, then try to reassemble something useful. In his opening comments to the staff, Farrar assured us that he would listen to our opinions before making any radical changes. He added that he would be less than honest if he failed to mention that certain changes were overdue.

Later that week, Dr. Farrar came to 111 and asked why I had requested (during the previous year) relief from the chairmanship of the Science Department. My answer prompted him to ask if I wished to comment further on the future of Ames High.

"Are you referring to modular scheduling?" I asked.

"That, or any other trend."

"When I plan a lab or discussion or test, it seems important to know that my students will be there. I understand that with modules you seldom know who will be present. I guess I'm not that flexible."

Farrar nodded, and I told him of my conversation with two biology teachers from St. Louis. We were attending NABT meetings in Philadelphia. The two teachers were reviewing the irregularities of their system, and I mentioned that our principal-elect was leaning toward small flexible modules. The two men gave each other knowing glances, and one of them remarked, "Undermine it at every opportunity!"

Although I am not a good miner - either over or under - I believe the principal knows my views and respects my stating them openly. Without actually shifting to modules, our schedule has become less rigid. Perhaps this was overdue.

We may have escaped becoming fully flexible because of another trend. Career education was ascending. There were grants for programs in which students were sent out of the classroom into the "real world."

It sometimes appears that Ralph Farrar is more concerned with his philosophies of education than with staff morale. And I suspect he would say this is as it should be: teachers have a good life. They can adjust. Student welfare comes first.

We could hardly argue with that last phrase. But what is best for the kids? Some of us subscribe fully to Farrar's views. Some are too private or too timid to speak out. Still others heckle, without fear of reprisal.

When improvements in facilities were on the agenda, we renewed our request for an enlarged greenhouse with a ventilated animal room. The request was filed, low on the priority list. One day at lunch, we hit upon a plan for underscoring the importance of the animal room. As chair of the department, Jones agreed to act. While Farrar was out for lunch, Jim took one of our cages of rats to the principal's office with a note: "May we use his space until the animal room is ready?"

He didn't give us the satisfaction of acknowledging our effort. Farrar is my seventh principal. It appears he will be my last. If more were to come, I suspect they would represent additional shades of democracy. My first, George Davis, used a psyching-out approach. He talked with conviction about an issue. Even if he was asking a question, you had the feeling he was answering it. By the time he sought our opinion, we knew how to vote.

Elmer Darling, the lanky Ph.D. who came from somewhere in the Dakotas, was with us only one year. I cannot remember voting. One of our assembly programs, however, is vivid. A traveling cast presented The Taming of the Shrew. At the end of the second act, Elmer came on stage, thanked the cast for a fine performance, and dismissed the students. Someone shouted, "There's more! We want to see it all." We stayed.

J.C. Wright was a skillful manipulator of teachers. Of students, too. He had been my chemistry teacher and track coach at KHS. When I returned as a teacher, he was sometimes my partner in ping-pong doubles. With Darling's resignation, Wright became principal, then superintendent. And after I had moved to Ames, he was appointed State Superintendent of Schools.

With Wright as principal, our staff meetings were well organized, including approval of minutes. Yet there was ample wisecracking and goading to get on with the vote. I don't remember how I voted, but I suspect the real urgency was to get on to the ping-pong tables.

Bob Laughridge was next, having moved up from the junior high school. Although Wright vacated the principal's office, he did not relinquish authority. It appeared that he hassled his successor. I recall the walk-out by a group of students, mostly athletes and cheerleaders. They were protesting Bob's discipline of two basketball players. When he announced temporary suspension of those who had left school, Wright took command. He called an assembly of all students, told them not to do it again, and reinstated all offenders. Laughridge moved to a position in Bloomfield the following year.

Vern Young, the head of Ames High when I arrived, was the most openly autocratic of my leaders. He was short, erect, and his shoes glistened. I was told that he kept a shining kit in his office and used it as occasion demanded. Each Friday morning of a home athletic event, Young assembled the male teachers on the lower bleachers of the gym. Standing below us with a clipboard of papers, he issued directives to ticket-takers, ushers, and guards - seldom asking for comments. All teachers were required to turn in Plan Books before leaving the building Friday afternoons. Our plans were to be complete for the following week. Mine were rejected on one occasion because they were written in pencil rather than ink.

Students called our principal The King.

While I was on leave with the Navy, Vern abdicated in late August. It was rumored that he started a farming venture in Colorado, and I wondered if he took his shoe-care kit.

Many years later, Young returned to Ames and attended a class reunion. As part of the program, former students presented him with a crown, made of foil-covered cardboard. He wore it with dignity while reviewing some of the highlights of his reign.

Herbert Adams accepted the uncomfortable task of organizing a high school in less than a week. Returning to Ames High in the fall of 1946, I found that our principal was dedicated to a "comprehensive" program. There should be courses to meet the needs of all. Activities were a vital part of school life, and my share was to sponsor the Spirit, Ames High's yearbook. When I asked whether this assignment might be closer to the English Department than to science, I got an impression that was borne out in the years that followed: There were certain issues on which there was no need to vote. A difference of opinion could be settled by a lengthy conference. Often the conference came in installments. Herb appeared to feel a strong obligation to clarify his position. He concluded that my interest in photography provided a fine opportunity to serve the students of Ames High School.

Our staff meetings included moments of hilarity. I recall one meeting that began in a dull tone but shifted abruptly when Herb said more than he intended. He was urging membership in the National Education Association. At one time, membership in both the national and state associations was routine. Now teachers were feeling more independent; there was talk about unions. Herb was assuring us that NEA membership was optional. "You don't have to join. You all know that it is a personal matter. It is a way of keeping up and working together as a profession. It is not required, you understand; but we all know that it is nice to be 100%. It is your personal decision. Just last year one of our social science teachers elected not to join. That was her right. Those of you who are new to the staff won't know her because she isn't with us this year."

As laughter mounted, Herb tried to explain that Karen had left to continue graduate work. But it was hopeless. Every time he started another sentence, spasms of laughter erupted. He laughed, too.

If an administrative policy was challenged, however, or if the high status of Ames High was questioned, Herb became defensively tense. There were no jokes. During Herb Adams' last years at Ames High, he was handicapped by the threat of another heart attack. Vice-principal Everett Ritland assumed more administrative duties. Still, Herb expended enormous energy defending the fine traditions of Ames High, attempting to preserve the old unity in a period of unrest and dissension. I suspect that every suggestion of change depleted his vitality.

Picking on principals is perhaps a sort of therapy for teachers who are picked on by parents. Yes, Ralph, Herb, Vern - and any others within reach - it is unfair to characterize an administrator in a few paragraphs. You have a complex task. I am glad I am a biology teacher.

Lockout

We were locked out of our hall of learning! During the night, vandals closed the steel gates at both ends of the science wing, securing them with heavy padlocks. Perhaps they were only pranksters; vandals would have entered our labs with axes and chopped our facilities. When the custodians failed to find a way in, Paul Olsan arrived from the metal shop with hacksaws and bolt clippers. While he worked, a cluster of students hovered in the background, watching the clock and smiling. When the gate opened, there was a mixture of groans and cheers.

Do the groaners know that there are more elegant ways of locking themselves away from opportunity?

Big Fleas Have Little Fleas

One way to kill a pesty caterpillar is to sprinkle it with a pest of its own. Some of the Lepidoptera larvae are among our most destructive insect pests. European corn borer, gypsy moth, cabbage worm. The list is long. And because many of the chemical pesticides contaminate the environment, control of such pests is difficult.

That is why entomologists are so interested in a little microbe that makes its living in the gut of Lepidoptera larvae. Bacillus thuringiensis is its name, B.t. for short, and it is a double threat to caterpillars. When the microbe runs out of food, it becomes a spore. It can live for years in this dormant condition. In becoming a spore, it produces a toxic crystal of protein. Billions of spores, along with the crystals, can be mixed with an inert carrier and sprinkled (or sprayed) on the food of the pest.

When a caterpillar eats this biological pesticide, it loses its appetite, turns black, and dies.

Other forms of life, including nearly all other insects, are not harmed by B.t. This is one reason why entomologists are so busy testing B.t. techniques on corn, cotton, cabbage, and other plants that are bothered by caterpillars. The Agricultural Research Service and Iowa State University, for example, are working on natural controls of the European corn borer. The microscopic bacterium is one of their weapons.

Could we demonstrate use of the microbial pesticide in the school laboratory? Encouraged by the ARS, I decided to try.

Before involving students, I tested some of the ARS techniques, rearing the corn borers on a synthetic diet. But the borers didn't cooperate as well at Ames High School as they did in the research laboratories. I tried growing corn seedlings at our south windows, inoculating the plants with corn borer eggs. Again, the results were not convincing. When I told Dr. Leslie Lewis of my difficulties, he suggested that I switch to wax moths.

Nasco came to my rescue with a culture of wax moth larvae. They arrived in a plastic tube, eating a mixture of baby cereal and a few supplements.

I had confronted wax moths occasionally in our beekeeping projects. The adult moths lay eggs on old combs, especially those in storage and those in weak colonies where the bees are less vigilant. The tiny larvae burrow through the combs, spinning silken tunnels as they go. Wax moths don't appeal to beekeepers, and I had never guessed that some day I would encourage them to live in my laboratory.

Now that a reliable host species was available, I involved some of the students. We discussed questions that we might be able to answer about the use of B.t. Does it really control the wax moth? Is I temperature a factor in its effectiveness? Is the age of the larvae a factor?

With plenty of larvae available, we transferred some to old crumbled honeybee combs, and tested their response to the insecticide. We thought the effects might be different when the insects were on a natural diet than when living on a synthetic food.

Our source of spores and crystals was Dipel, one of several commercial preparations of B.t. We placed 100-gram samples of food in petri dishes and mixed into it varying dosages of the insecticide-1 gram, 2 grams, 4 grams, and 8 grams. Then we added 6 medium-size larvae to each dish. In each test, one dish without Dipel served as a control.

For testing the age factor, we sorted out two size groups of larvae, about 6 mm long and about 12 mm long. For the temperature test, we kept one set of dishes at room temperature, about 20 degrees C., and another in an incubator at 30 degrees C.

We found that five days of observation were adequate after a test was set up. Before we had finished, another generation of larvae had appeared in the pickle jars. In fact, we were able to donate a few fat larvae to students who wanted to try them as bait for ice-fishing.

The test results? For one thing, the fish were biting, and the catch was good. For another, wax moth larvae are clearly vulnerable to Bacillus thuringiensis. Although the testing will probably be smoother next year, both diets worked, and the kill was convincing. Confusion in our records made the temperature test of doubtful value. We found that the older larvae were more vulnerable than the younger ones, and 2 grams of Dipel in 100 grams of food appeared to be enough. We assume that the optimum dosage might differ for other brands of the insecticide.

Some of the questions that came up during the investigation may have been as valuable as the experimental results. Why is the parasite of an insect unlikely to be dangerous to humans? Why are alien insects more likely to be destructive than native species? When control specialists talk about LD50, what do they mean? (They refer to a "lethal dose" that kills 50% of the subjects tested). Why don't bees sting wax moths that try to invade their hive? (They often do). Can we open some of the wax moth cocoons and see what's going on in there? (Sure! Help yourself to understanding. That's one of the good things about wax moths!)

> So, naturalists observe, a flea
> Hath smaller fleas that on him prey;
> And these have smaller still to bite 'em;
> And so proceed ad infinitum.
> Thus every poet, in his kind,
> Is bit by him that comes behind.
> ..Jonathan Swift, 1733

REACHING OUT

In our staff meeting today, Farrar was urging us to set clear goals for improved teaching. He calls them goals for growth, and he cited Dave Posegate's experience. Dave had tried many diet plans in an attempt to lose weight. All plans failed until Dave set a definite goal-to lose 20 pounds during the school year. Knowing exactly what he wanted to accomplish, Dave did it.

This prompted Barb Ward to comment, "I hardly see how you can call losing 20 pounds growth."

Undaunted, Farrar said we should not set a goal that we would normally accomplish anyway. It should go beyond normal expectations for growth. It should require reaching out.

Rising with the success of her previous quip, Barb asked for an example of reaching out.

"Well," said Farrar, "I think it's a good idea to start each week with a written plan, so you know where you're going. Written plans generally improve teaching."

"I've been doing that," returned Barb. "Would it be growth if I started doing what's in my plans?"

After the applause subsided, Ritland called out, "The score is now two- zero!"

Peppermint Ice Cream

Now I'll know what's wrong if my car stalls when I buy peppermint ice cream. I learned from the Time-Life Book of the Family Car. It is a fine lesson in biology.

In physics, too. And in economics, international relations, sales psychology, geology . . .

The story concerns a lady who complained that every time she bought peppermint ice cream, her car failed to start. No trouble with vanilla, just peppermint. The service manager who listened to this complaint probably thought he should apply for a refresher course at Detroit. But he listened. And to make sure that he himself was properly tuned, he accompanied his distraught customer on a test run. They stopped at a dairy store, and the lady bought a pint of vanilla ice cream. When she came out, the car started all right.

Then, after driving around the block a few times, they stopped again, this time for peppermint. And just as she had described, the engine balked. (By the time I got that far in the story, I was glad that I was a biology teacher, not a car mechanic!)

Well, the service manager survived the crisis. He recalled that when his client bought peppermint ice cream, she was in the store longer than when she bought vanilla. He found that while the vanilla flavor was pre-packaged, peppermint was a bulk item. "While the woman waited a few extra minutes for the peppermint ice cream to be hand-packed, the fuel in the car's carburetor percolated from the bowl into the engine, flooding it. This did not happen with vanilla ice cream because the woman came out of the store before the problem developed..."

The biology lesson? After relating the ice cream story in each of my classes, I looked around the room expectantly, hoping someone would say something useful. Mostly, they stared at me as though waiting for a punch line. But, in one class, a girl said something useful. Almost accusingly, she remarked, "I suppose you want us to think about an experiment, maybe about good controls."

Exactly. I wanted the students to see that when two events occur together, it doesn't always mean that one is a cause, the other an effect.

One reason for the "success" of so many arthritis treatments is that arthritis patients commonly have periods when the stiffness and pain disappear. No one knows why. But if the patient has taken a new pill, a vibration therapy, or a shot of bee venom, it is easy to attribute the remission to the treatment. There could be a relationship. But there may be none at all.

I once tested my students with a think question concerning the relationship between school grades and smoking. I cited a hypothetical situation showing that in two groups of students, 500 in each group, the average grade point was significantly higher for the non-smokers. The students were to interpret this finding.

In the discussion that followed, students tended to take sides, those defending smoking versus those against it. John acted as though he had settled the argument by announcing that smoking hadn't kept his dad from becoming a DVM and a Ph.D.

I am not sure that I ever convinced some of the students that we were not a jury. We didn't have to decide whether smoking causes low grades. We were merely asking ourselves whether there was enough evidence to decide why 500 smokers had a lower grade point. There wasn't.

It was useful, of course, to think of possible hypotheses. One student bravely said that teachers are probably prejudiced, that smokers don't get a fair shake at grading time. The most obvious hypothesis was the one that many students thought we should settle, that smoking causes low grades. None of my 116 sophomores, however, suggested an equally interesting explanation, that low grades cause smoking.

If I had just known about peppermint ice cream!

My Unreal World

Many years ago, when job experience became a credit subject, only students who had no hope of formal education beyond high school were released for on-the-job training. Applications were carefully screened. One of the shop teachers supervised the program, placing the students in appropriate jobs and conferring with employers. It was a good program.

It is still a good program.

It is good for certain students. But there is a built-in danger of involving students who should spend more time in the "unreal" world learning to read. The employer obviously prefers a bright student. And the school looks better if it sends a bright student.

A recent recruitment brochure was headed, "What school did you learn in today?" And the final paragraph asserted, "Because of the real-life problems students must learn to deal with, the important question is not 'What did you learn in school,' but what school did you learn it in?"

I suspect some of the students who should be learning chemistry or Spanish are learning to pump gasoline or to operate a hair dryer. They are being short-changed right along with those who need work experience most but fail to find a place in it.

Martin, I'm Listening!

Martin is the small boy who comes in talking. He goes out talking, too, and his speech reminds me of a sputtering 20 mm. machine gun.

The first burst is likely to come as he walks through the door, a minute or so before any of the other students. During class, he volunteers miscellaneous bits of information, raising his hand for permission just after he starts firing. There is often an unfinished comment at the end of the period. He starts on this as he rises from his seat, talks his way to the front of the room, and stops at my desk. From there he has better aim.

Being far brighter than most of the students, and exposing his intelligence freely, Martin is the butt of frequent pranks. He sits on a wet paper towel, for example. His books disappear while he is checking on the welfare of the tarantula. As reports are passed forward, his is mislaid.

This is how it worked out today: Martin came forward to ask if I had been teaching full-time while writing Foundations of Life Science. After hearing my answer, he returned to his seat, only to come forward again and report that his notebook had disappeared.

"Those things usually show up," I said and walked back to check the empty seats. Because there were no furtive glances following me, I assumed that Martin's usual hecklers were innocent. Suspecting that the notebook would mysteriously reappear, I returned to my work.

Martin followed me forward. "May I go call the police?"

"No, Martin. That wouldn't solve your problem."

"It's a valuable notebook. I wouldn't be able to retrieve all of the information in it."

"I'll try to find it."

"May I come and get it at 3:20?"

"I'll try to find it. Come if you wish."

Fortunately, my next period was free. After searching the bookshelf of each lab table, I opened all of the storage compartments that had not been locked. Having exhausted the obvious hiding places, I checked the courtyard. Perhaps someone had thrown Martin's retrieval system out the window. Next, I checked class schedules of students who work near Martin, and spoke with those who were accessible. Each of them appeared to know my mission and quickly denied taking the notebook.

"I didn't ask if you took it," I said. "I'm asking if you can tell me where to look for it." - no clues. Near the end of the period, I checked Martin's schedule and waited outside the computer room for him to appear. When the other students had left, I found Martin at Dale Hiedeman's desk, firing away. I walked in and stood beside Martin. He expended a few more rounds, then gave me a fleeting glance and said, "Oh, Mr. Trump, I found my biology notebook in the math room."

"You had math just before biology, didn't you, Martin?"

"Yes, but I think I took it with me to biology."

The crisis over, I returned to the serenity of my eighth period biology class.

Martin surrounded me with words. After I asked him to slow down and begin again, he said, "I just want to congratulate you on what you said in your book. Most people call it a microphotograph. You called it a photomicrograph. You are right. A microphotograph is a little picture of something. It could be a little picture of something big. A photomicrograph is a big picture of something microscopic. Not many people seem to know the difference between a microphotograph and a photomicrograph. You have it right."

"Thank you, Martin," I said. "I'm relieved to know it's right. Keep me informed."

Martin's notebook again! He pushed his way among his classmates at the end of the period and stood before me with that shall-I-call-the-police look. This time I had an answer. "Martin, your notebook is right here on the stack. You passed it forward with the others at the beginning of the period. You may retrieve it tomorrow."

When Martin moved in on me today, I knew there was a real problem. His ten fingers were interlaced, sliding back and forth in a pumping action. This is what he said: "You know that idea about the ants and the acacias in Australia and your book says there aren't any hollow spines on those trees maybe because there weren't any browsers there when the plants and the ants were evolving and it says this shows it is mutualism but I think all it shows is that acacias don't really need any ants to protect them because the spines are enough to do it without any ants, so I think maybe that entomologist was right and it isn't mutalism after all what do you think?"

Lucky for me the class period was over and another thirty students were coming in and I was glad because Martin may have a point and that biologist at Harvard University is wrong and I'd better read pages 210 and 211 again someday because maybe it's not really mutualism. Which would indicate that I'd better avoid such controversies.

Martin, the fifth period just wouldn't be the same without you, but I'm glad you are one of a kind.

Goals By Admonition

Again, we are admonished to put our teaching goals in writing. And so I have committed my energies. I will: provide frequent practice in distinguishing between fact and opinion; help students become more wary of quackery; encourage satisfaction in reading such magazines as National Geographic and Today's Health, as opposed to such as Mad; help students learn about political candidates' views on environmental problems; encourage tolerance for inherited variations within the human species, greater respect for individuality; encourage tolerance for persons who believe in the evolution of life; encourage tolerance for persons who do not believe in the evolution of life; help students avoid being intimidated by scientific terms that appear in news media; and help them evaluate sources of information, with caution about authoritarian views, including mine.

I would feel encouraged by these commitments if it were not for one statement that I found while proofing my list. It followed a goal for helping students write and speak so they cannot be misunderstood.

"Encourage students to have no more than two children, doing so indirectly rather than by admonition."

Neural Speedway

Many of the lessons that I once thought indispensable are no longer in my plan book. One of the survivors is a way of measuring human reaction time that I found in Current Science more than 30 years ago. It requires two long pendulums, a clock, and a bunch of competitive kids. They like it, and I like it.

The question: How long does it take for a message to go from your eyes to your brain, then down to your fingers and evoke a response?

Our pendulums, one slightly longer than the other, are made of string, with lead weights, suspended from a horizontal bar above the demonstration table. The student being tested holds the shorter pendulum back so it can swing freely when released. The tester holds the long one in a similar position. Without warning, the tester drops the long pendulum, and as quickly as possible the testee drops the short one.

The long pendulum has a head start, but the shorter one swings slightly faster. The number of swings for the short one to catch up is a measure of reaction time.

In our arrangement, the long pendulum completes 33 cycles (forth and back) in one minute, a rate of 1.82 seconds per cycle. If we start the two swinging together, the shorter one is soon ahead and eventually gains a full cycle. It does this in 112 cycles. Dividing 1.82 by 112, we find that the short pendulum gains on the long one by 0.016 second every cycle. A count of nine cycles, for example, indicates a reaction time of 0.144 second. I have read that in humans the neural impulse travels 100 meters per second. For a process that powers its way along the fibers by such complex chemistry, that's moving. It is ten times faster than a good sprinter sprints.

Since the distance from eyes to brain to fingers is less than one meter, reaction time must include a delay in the response of the muscles. Perhaps the message is also delayed in making the right synapse connections in the brain.

Whatever the process, reaction time varies greatly among my sophomores. Watching some of them multiply by 0.016 and search for a place to put the decimal point, I suspect there is no correlation between reaction time and ability to deal with numbers.

Unfinished Business

I left the woods for as good a reason as I went there. Perhaps it seemed to me that I had several lives to live, and could not spare any more time for that one.

Thoreau said it.

I have three more years to leave the woods voluntarily, and to explain why. But I won't leave voluntarily. Although I am not sure how many of my lives remain, there is a lot of unfinished business in this one. Shortly before I am 65, I will receive a nice letter from the superintendent advising me that I may wish to retire. It is mandatory.

When I consider the things I haven't completed, or even started, in 38 years, I feel a slight panic about the three that remain.

It would be interesting, for example, to teach for a year or two without assigning grades, only pass and fail. Yes, I know that the pass-fail system, as applied to individual students, hasn't worked well. Too many passers perform at the D level when they should be learning something. But suppose the entire class were on such a system. Would some of the members rediscover excitement in learning? Would they regain the kind of curiosity that keeps a child's mind working?

It would be satisfying also to find an answer to at least one of my questions about woodchucks. What is the meaning of the intolerance that we have observed between male and female? Is it a seasonal thing, typical of the species? Or does it apply only to intolerant individuals?

When we released the first courtyard chuck at the old high school, a mature female, she dug a den and accepted a sub-adult male that we released three weeks later. Before hibernation in late November, they excavated two additional dens and carried dry grass into them. Were they preparing to hibernate in separate dens? Were they anticipating the eviction of the male during a possible gestation period the following spring? Or were the extra dens merely alternate escape routes in a crisis?

In later years at the new building, we saw territorial disputes from the very start. On a couple of occasions, we installed the microphone of a tape recorder in an occupied den and released another woodchuck at the entrance. The recordings told of intermittent scuffling, whiny snarling, and tooth chattering. One pair denned at opposite ends of the courtyard.

By flooding the east den and blocking the entrance, we tried to force the two animals to share the west den. They persisted with separate lives.

It would require more than casual observations to interpret such events. Perhaps we should eliminate the study of insects and plant life in autumn, and genetics in spring. Then we could concentrate on my favorite rodent!

Perhaps in the next three years I should try admitting that there are stupid questions. I have maintained that any sincere question is valid. Attempting to draw out the minds of reluctant learners, I have insisted that asking questions is a way to learn. This is often true. But I must admit that there are questions and there are stupid questions. Perhaps I should try snarling at trivia.

And how about dealing with the student who returns after an absence and asks, "Did I miss anything important?" I wonder how it would affect both me and the student if I were to reach across the lab table and stretch his nose.

Speaking of my lab table, how would it be to teach for three years with an uncluttered desk and demonstration table? Though I would like to try, I am uncertain of the outcome.

The problem appears to be with my filing system. It looks as though it evolved through random mutations, without the benefit of natural selection. Yet there is a certain logic in my system. It started with a certain compartment for all of the beakers. Then I decided that it would be convenient to have a certain size beaker with the material needed for the study of diffusion, another size for the gear for hydroponics, still another for the size needed in testing the pH of Squaw Creek water. Eventually there was a beaker in nearly every storage compartment of the lab, a few microscope slides in every drawer. And so on. If there are twenty units of an item, they are found in twenty places.

This is not all bad. I can generally find one of them. The trouble is that one of those places is on top of my desk, another on the workspace of the demonstration table. A few years ago, when the authors of Creative Biology Teaching wanted a picture of a badly cluttered table, an untenable teaching situation, they sent a photographer to room 111, Ames High School.

If I could reorganize my filing system in two years, there would be one year left to study the effect on education.

While I am living in a dream: In these last three years I should find a way of knowing my students as well as I knew those of my first three years. During that period in Keokuk, I paid more attention to individuals. Even if I failed to impress them with the nature of life, I could see through their skulls more clearly and know what was happening inside.

Was my insight better because of the energy that I was forced to spend keeping my classes under control? I talked with small groups, with individuals. I questioned their motives, tried to interpret their behavior. Whatever the connection, it is sad to suspect that with increased knowledge, a teacher could become less personal.

Some years ago I hit upon a scheme for individual conferences during class time. Making arrangements with the librarian, I sent half of the class for reference work one day, the other half the next day. I talked individually with those who remained in 111. A part of the strategy was to send one student from each worktable. This meant that I could move around the room in an irregular path, stopping at each table, visiting with each student privately. Although the conversation sometimes turned to a specific lesson or problem, I tried to steer toward outside interests. This undoubtedly helps. Yet I suspect that it is too obvious. If I could gear my time to do this type of thing more often, more casually, it would be more valuable.

Could I squeeze in a year or two of self-paced biology? If you say this hyphenated term in the right company, you are in line for a promotion. The idea is that students progress at different rates. We've recognized that for years. We've given those who make fast progress A's, those who make no progress F's. Now we are supposed to allow students to move from one topic to another at their own speed. Like powdered pies, lessons are compressed and preserved in little modules. When a student is ready to move to the next topic, she or he checks out a module with instructions, references, gadgets, and perhaps with a taped greeting from the instructor. (I could have said that the student is issued both hardware and software.) Then he or she walks at an individualized pace to a semi-private spot and begins work when convenient. (If self-pacing had been invented a couple of generations ago, the semi-private place would have been called a study stall; now I believe the proper name is carrel.)

In courses like physics and math, involving inanimate paraphernalia, supplying the hardware and software would be reasonable. But with some students examining corn embryos, other identifying protozoans, still others checking their blood types-well, the procurement problem alone might drive me up the hall.

Teachers at Ankeny, Webster City, and Cedar Falls say that their students are thriving on the self-pacing plan. It would be nice to say I tried.

Also, for years I have assumed there was plenty of time for me to really get involved with a lesson on dissection. Many fine biologists were initiated on carcasses of frogs and dogfish sharks. Scalpels are still basic equipment in some high school labs. After giving dissection a prejudiced trial in Keokuk, I restricted it to a couple of specific adventures. We might observe how the ventral nerve cord of an earthworm encircles the forward end of the food tube and expands into a brain. Or we might expose the heart of an anesthetized frog and see how temperature affected the contractions. But we never dissected an animal to see how it was assembled or to learn the names of its innards.

Today, although I contend that dissection is seldom worth the wear on the scalpels, I find at times that I am expected to apologize for not having students mutilate the anatomy of a cat. If I were to try it again, the outcome might persuade me to recant. If not, my argument would be more convincing.

In matters of discipline, I have toyed with the thought of trying techniques that seem to work for my colleagues. Generally, I have respected the hands-off view. Spatcher, on the contrary, sometimes grabs a student by the hair to be sure that he is being heard. And Spratt massages a boy's shoulders, burrowing halfway through the trapezius muscles with his knuckles. Bob Impecoven, I believe, would not interfere with a fight. "Let them knock themselves out," he would say. "It will save me the trouble." I recall one instance when a boy smuggled liquor to his locker and imbibed too much before going to the study hall. When the boy became sick, Bob would not allow him to go to the rest room and clean up. He sat in the midst of the vomit.

This could go on and on. But it won't solve my woodchuck problem or clear my desk for tomorrow.

It's Not a Federal Offense!

My running battle with my homeroom students gained momentum this morning. Bert was late again.

Two weeks ago when I remarked to the group that promptness counts heavily with employers, several students argued that this just isn't so. Jim, whose record is good, agreed with two other boys. They said their boss didn't object if they were a little late.

Attempting to resolve this part of the problem, I called the Iowa Employment Service, obtained approval to record the conversation, and asked about the attitude of employers to punctuality. Chuck Savage strongly affirmed that it pays to be on time. So did Dale Tramp, one of our counselors. Dale said that when employers call for recommendations, the tardy and attendance records are their first concerns.

Having these comments on tape, I offered to call the employers who don't object to late arrivals. I said I would tape their comments along with the others. There were no offers of employers' names.

I thought this resolved the question. But Bert came in this morning after the tardy signal. When he found that I had listed his name, he said, "How come? I was in the room 30 seconds after the bell!" As he left for the next class, I tried to ask at what point he thought I should draw the line.

He replied, "No, Mr. Trump. You're screwing me." And he walked out.

No arguments in homeroom this morning. I remarked that if anyone wanted to remake the rules, he'd have to talk about it – if not with me, then with one of the administrators. I aimed directly at Bert. Because he had challenged me in front of other students, I thought I should accept the challenge in public. Bert nodded, and there were no further comments.

Later in the morning, I learned from Bill Ripp that Bert had been in the office requesting a change of homerooms. Bill said no.

At noon, I had the good fortune of heading for the office area just as Bert started down the hall in that direction. He had no choice but to walk with me.

"What do I do, Bert, that makes you think I'm picking on you?"

"I don't think that at all. You're unfair to all of us."

"How?"

"You make a big deal over tardies. It's not a federal offense."

"Find out what kind of offense it really is, and let me know how to decide who's late and who isn't."

"I'm on my way to see Mr. Ritland now."

Later that afternoon, Rit told me that Bert and Dan had come in with their complaint. He told them I was following school policy and if they had any real problems to come back. They shrugged and left.

===

Bert is the winner. A note from the office says he has accepted a part-time job that will interfere with attendance in home-room.

He didn't ask me for a letter of recommendation.

TREE-SITTER

Jan Eldredge is back from the wetlands of the north with a fine set of wildlife photographs. There is a mixture of elation and jealousy when I find one of my recent students succeeding where I have not.

I remember telling classes of my difficulties getting deer to trigger my camera traps. Jan came in after school with a plan. If she could borrow one of my cameras, she would go upstream on Skunk River and sit in a tree at night. When a deer came by, she would photograph it for me. The following Monday she reported, "I sat there a long time. The deer didn't come."

Now that she has a degree in wildlife from Iowa State, Jan sits in a canoe instead of a tree. If the ducks and otters don't come to her, she goes to them.

A GROWTH OF SOUR GRAPES

Perhaps I will never recover from the belief that I am about to begin writing a textbook of biology. When I see a wasp homing in on its hole, carrying a cricket, when I find a patch of lichen pioneering on bare rock, when I read of a mutant microbe that is tolerating a potent antibiotic- finding such bits of biology, I store them automatically in my head, hoping they will be there when needed.

I should have been cured by sweating through Design for Life. This project started in 1960 after Holt salesman Ray Hayes asked my opinion of Modern Biology. I explained why I thought it encouraged memorizing definitions rather than understanding organisms. Ray said he'd like to forward my ideas to the editorial department.

A month or so later, chief editor Milton Hopkins wrote asking if I would meet him in Des Moines to discuss authorship of a biology text.

The result was Design for Life, a second-year text that was geared to Advanced Placement standards. Holt was betting on a surge of interest in advanced biology for high schools. Holt was wrong. The book was used in a few aggressive places such as the Bronx High School of Science, and in an elementary course at Purdue University. Before that fate became obvious, Holt decided that because of BSCS impact, Modern Biology was doomed. Would I consider writing a book with Design viewpoints for a beginning course? Did I have any suggestions for a collaborator? When could I come to New York for a conference?

That was the beginning of a seven-year fiasco. Holt was purchased by CBS. Milton Hopkins retired. Virginia Johnson, who had done a fine editorial job on Design, left Holt to join Addison Wesley. A succession of three editors followed, each less competent than the predecessor. There were long gaps when no editor was responsible for the project. Correspondence virtually stopped. The title of our book was changed from The Nature of Life to Foundations of Life Science. Holt believed a third author should be involved. They selected a dud, then dropped him after Roger Volker and I found him an impediment. Bruce Kroog, the final editor worked hard, beginning every letter with, "Enclose find..." What we found seldom lifted our morale.

As we closed in on the final chapters, color photographs (on which I had expended both time and material) were deleted. The cover shot that I submitted - a black-and-white of three gulls in a pleasing flight formation -was made into a sample cover but finally rejected. The artists decided it didn't look right - they had printed the gulls green.

When the first copy arrived, after many delays, Roger and I quickly discovered 31 errors. For example, the picture of the orb-weaving spider was turned 90 degrees. Those of the earthworm and human jawbone were upside down. A diagram of structural formulas was replaced by one that had appeared on a previous page, so the caption made no sense.

To illustrate the effects of rickets, we had selected a front view of a little naked boy. By the time it reached the presses, artists had performed surgery on the print, skillfully amputating the genitals. As if there weren't enough defects showing!

Such errors, however, were no great problem as far as sales were concerned. Our Iowa agent was the only one who appeared to know that the book existed. He retired soon after its publication. There were no advertisements, no review copies for the journals, no royalties.

In a magnificent gesture, Holt donated a complete set of books to Ames High. It was a relief to burn the stacks of duplicated sheets that had served as our text for two years. And now that my name is on the cover, parents are more keenly aware of whom to blame if a biology grade brings down the family record.

SETTLING IT

Several times during the second period, Jim turned around and glared at Wayne. At the end of the period, they were exchanging words. And after the bell, with a table between them, they were nudging and pushing each other. Although there was more ferocity in their faces than in their arms, fists were ready.

I told them I wanted to ask a question before there was any settlement. "Sit down," I said..."No. At the same table."

They sat, glaring straight ahead.

"What's the problem?"

"He's been kicking my chair," said Jim. "He's been kicking it for three days."

"Is that true, Wayne?"

"Yes."

"What's Jim been doing to you?"

"Nothing."

"Okay," I said. "I'm glad it's settled. You can go now."

Both looked startled. They walked down the hall, a table-width apart.

CHEATING IN SCHOOLWORK? "IT'S ABHONSIBLE!"

What do sophomores think about dishonesty in school work?

"If you plan to cheat all your life, cheating in school is a good way to practice." "It doesn't bother me if I do it, just so nobody else does." "Being a D student, I don't have many copiers. If they do, it's only once." "It is a waste of time for everyone, even the teacher."... "Deliberate cheating on a test? Well, it's abhonsible!"

Those are samples. Annoyed by the glib attitude that is common about dishonesty in schoolwork, I conducted a survey in our twelve biology classes. At the top of a half sheet of typing paper, I duplicated this request and distributed the papers without comment:

YOUR OPINION PLEASE!

What do you think about cheating in schoolwork?

(This is not concerned with any faculty project. I am merely interested in understanding student problems and opinions).

DO NOT SIGN YOUR NAME.

This started five years ago. Wondering how the Watergate affair and other public scandals might affect student opinions, I continued with four additional sets of classes. This is the last. It has been difficult to place all of the replies into categories. Yet, the replies to this open question are surely more revealing than would have come from a formal questionnaire.

About 8% of the students said they approved of cheating, while 35% said they disapproved. Twelve percent admitted that although they disapproved, they still cheated. Nineteen percent wrote that cheating hurts only the person who cheats.

A minor number commented on reasons for cheating. Five percent blamed the importance of grades in getting the right job or in being admitted to the right college. About equal numbers (3%) blamed laziness, pressure from the family, and the prevalence of cheating by others. One percent said students cheat because the lessons are not worth learning.

More students commented on reasons for not cheating. About 12% took the view that it is unfair to those who are honest. About 2% cited fear of being caught, while another 2% said that the feeling of guilt is too strong.

What should the teacher do about cheating? Several students mentioned an honor system but seemed doubtful about how it works. Five percent recommended that teachers do what they can to prevent cheating, while another 2% said teachers should ignore it.

That is what 1740 Ames High sophomores said. Is that what they believe? One of them said that cheating is bad for everyone, then added, "I'm not saying this because I know that's what you want me to say. It's true!" But what of the other 1739?

A month after the open-response survey, I used a more objective questionnaire, without signatures. It contained all of the categories that I had previously gleaned. These were the first three questions, and the means of the replies:

In your opinion, what percentage of Ames High students are honest in their schoolwork? 57%

In your opinion, what percentage cheat occasionally? 48%

In your opinion, what percentage cheat regularly when this is possible? 24%

The final question was this:

What is your own record of honesty in schoolwork? Check one.

Do not cheat 45%

Cheat occasionally 53%

Cheat regularly when this is possible. 2%

I am puzzled by two of the figures in this survey. If 24% of Ames High students cheat when possible, how is it that only 2% of the reporting students do so? Do biology students believe they are that much better than their peers? Self-criticism comes hard.

The greater part of the objective questionnaire asked for judgments about the relative importance of various ideas. For example, do students cheat because lessons are not worth learning? This factor was considered only slightly important. It received a rating of 1.1, zero being unimportant, 3 being highly important. And does cheating really hurt only the person who cheats? There was moderate agreement that it does, with a score of 1.9 on the 3-point scale.

I've been especially interested in this myth, that cheating hurts only the cheater. It came up often in the open responses. Perhaps parents and teachers perpetuate the idea, believing that it discourages cheating. Dishonesty hurts everyone. It erodes morale.

As an economist might say, it is counterproductive. The myth also leads students into another dead end. If only the cheater is hurt, then the teacher should ignore cheating. That's downright depressing!

Did sophomore opinions change during the five years of sampling? Not much. I saw no need to add to the list of categories that I made during the first year. And some of the percentages were remarkably uniform throughout the surveys. For example, the five-year average for students who volunteered that they disapproved of cheating was 12%. The low was 10%, the high 14%. The average for those saying that punishment for cheating should be more severe was 4%, with a low of 2% and a high of 6%. This was not a distinct trend. The high years were the second and last.

Were student opinions altered by the publicity on dishonesty in public life? There was a trend of opinion regarding three categories: the idea that everyone cheats to some extent, blaming the importance of grades as a cause of cheating, and the assertion that cheating is unfair to those who are honest. There were fewer students expressing these three ideas as the survey continued.

One category appears closely related to the level of honesty that students observe in adults as well as in their peers. An average of 3% said that students cheat because there is so much cheating in everyday life. As one student remarked, "The system requires it." The sequence of counts that appeared to fit this category was 4%, 4%, 1%, 4%, and 1%. And from conversation with social science teachers, I suspect that not many of our sophomores were concerned about the high level scandals that were eroding our confidence in government.

Now, what to do? I have shared my ideas and data with Dr. Farrar and a few teachers. Once it was suggested that I would be asked to report to the entire staff. Nothing came of it.

I would like an informal exchange of ideas, followed by a quiet crusade. No announcements, no student council advice, no speeches. If our only action should be to watch for subtle ways of showing that we care, it would be worth the trouble.

I would like more people to know how a Canadian teacher handled a plagiarism problem. A talented boy who had been writing sports stories for a local paper turned in a poem copied from the Saturday Evening Post. The teacher scored the assignment with a single comment: "This is not up to your usual standards." The boy understood, and asked if he could submit something else. The second poem, I assume, was not Post caliber, but the teacher remarked that it was done in the boy's "unmistakable style." (The account is part of an article by R.G. Martin in English Journal, May 1971. John Forssman spotted it and passed his copy along to me.)

The least we could do is to remove some of the opportunities for cheating. This would be one way of showing that we care.

There is satisfaction merely in looking over the sample comments that I have saved - the witty ones, the sad ones, the inscrutable ones:

"I don't see all that much harm in copying a few answers from your neighbors. After all, they could be wrong." "And yet if you know the material and your mind goes blank, what else can you do?" "I respect myself too much to cheat, no matter how easy it is." "If I don't give answers, my friends hassle me. I don't like being hassled." "It's terribly easy to cheat, and it takes a lot of self-discipline to yield to temptation." "I thought it was cool the way you rearranged your test questions so they weren't the same for people sitting together." "But there are limits. I don't think you should look at your neighbor's cheat sheet if you are too lazy to have one of your own."

"I don't like people copying me. They come around and blame me if it's wrong." "It tells that you're not proud of yourself and you do stuff to make you look better and if you have friends, they won't be proud of you either." "Anyway if someone asks me for answers, I feel flattered because he is depending on me." "I tried cheating a few times. It isn't worth the eyestrain." "Some parents threaten their kids, like cutting their allowance or not letting them use the telephone. One of my friends got his stereo taken away. I feel sorry for kids with parents who think everyone can get good grades." "You asked what I think. Well, I think I copied this from the person in front of me." "Give him the grade he got by cheating. It will serve him right!"

On and on! I regret that I do not know the authors. Names might help me recognize which of the responses are honest. And perhaps I would learn the meaning of abhonsible.

It Was A Lie

Harold is tall, Beth slightly taller. He leans on her, his arm around her neck. But I suspect things are not going well with them. They passed me in the hall as I was coming out of 106. She was shaking Harold off her neck and saying, "Yes, that's what I said yesterday. But it was a lie. Every nice thing I've ever said about you was a lie!"

There, Harold, take that. And get off that girl's neck!

Are There Any Questions?

Farrar: Before we get underway with the agenda, I want to mention a matter relative to Parent Visitation Night. As you know, this will be Thursday evening. Parents will go from class to class on a ten minute schedule. They'll follow their offspring's Wednesday schedule.

Voice from second row: Was that the Wednesday schedule on Thursday or the Thursday schedule on Wednesday?

Farrar: The parents will be here Thursday night. It will be helpful for you to come the same night. And I want to say relative to the student schedules, you'll have the schedule forms in your boxes Wednesday morning. Ask your students to fill out both halves. Students will take one half home. You'll bring the other half to the office. That way we'll have a copy for the parents whose kids didn't bring one home.

Voice: Alphabetized?

Farrar: Be sure to take care of that on Wednesday so you can catch Wednesday's absentees on Thursday. B comes right after A. C comes next.

Impecoven: We have a short homeroom on Wednesdays. Can we do it in five minutes?

Ripp: We have a ten minute homeroom on Wednesdays.

Voice: When my students arrive, there are only five minutes left, and checking attendance takes one of them.

Ward: Could we do it on Tuesday?

Another voice: Make out Wednesday's schedule on Tuesday, for use on Thursday? My students can't do that in ten minutes!

Far right: We won't solve the tardy problem till we all crack down.

Another: Teachers don't crack down. They crack up.

Ward: Let's do it Tuesday.

Voice: My students aren't all here on Wednesdays. Now that we're shortening periods on Wednesday, they don't all come.

Farrar: As I recall, we're considering Tuesday. But incidentally, how are we coming with calls to parents? That would be a good goal for growth; more contacts with parents.

Bailey: They can contact us Thursday night. Or is it Wednesday?

Farrar: Maybe we've already cracked! But let's agree on Tuesday. We'll fill out schedules on Tuesday, with make-up for Tuesday's absentees on Wednesday when not everyone comes. Where's the agenda?

Ripp: If we did it on Thursday, I'd have another day of grace.

Farrar: Grace who?

Bailey: May I review the plan? Ripp will make out the forms on Monday so we can...

Farrar: No, Keith. If you don't understand the plan, ask your students for help. (Ripp hands him a copy of the agenda). Yes, thank you, Bill. Now we have several items relative to education at Ames High School. . .

SUBSTITUTE FOR FIGHTING

We've been dealing with animal behavior. In fighting between individuals of the same species, there are seldom serious injuries. In fact, many species have found substitutes for real fighting. Two lizards will walk around each other stiff-legged, trying to stare each other down. Shrews engage in screaming contests. Such ceremonial fighting enables individuals to protect their territory without the hazards of battle.

A wild idea came when I remembered William James' Moral Equivalent of War. In 1910, he argued that man could divert his fighting from war to useful purposes. The same hardships could be applied to public welfare. Youths, he said, could be drafted for this work "to get the childishness knocked out of them, and to come back into society with healthier sympathies and soberer ideas."

As a departure from our study of animals, I told the students of James' idea and asked if they could see any parallel between the moral equivalent of war and the harmless "fighting" of lizards and shrews. There were bleak silences. In one of my five classes, a boy nodded and said, "I guess they're a little alike. I mean working on a welfare job wouldn't hurt like getting shot."

By the time my last class left the room, I wondered if I saw any parallel.

FILE 7

I have talked with biology teachers who were frustrated by restrictions in what and how they teach. In 41 years, I have had no such complaint. In fact, I have been encouraged to teach as I think best. My main frustration has been that I don't teach as well as I know how.

This admission struck me as I flipped through pages of File 7. Why this folder was labeled 7, I don't know. For most of my career, however, it has been a depository for an occasional memo issued by an administrator. They were messages that once frightened me, irritated me, or amused me. The earliest issue that I found was one placed in my mailbox early in World War II, when I was still in Keokuk. The full sheet contained only two sentences: "In view of the paper shortage, please do all you can to conserve. The superintendent has notified us that the present stock must last through the current school year."

My reason for sifting file 7 was to discover whether there was anything that I should pass on to my successor, or take home as a personal treasure. Many of the papers were difficult to classify. An ancient installment of advisory council minutes, for example, had only one sentence that appeared important enough for filing: "Trump moved the meeting adjourn." Was I that fond of seeing my own name?

A full record of the various improvement committees would reveal a persistent threat to the tranquility of Ames High. It was the sound system, used mainly during the homeroom period each morning. Student organizations as well as staff members were granted time to broadcast information throughout the building. The volume, both in number of words and in decibels, was often excessive.

Even worse, there were occasional emergencies that blasted into our classrooms later in the day. The Drama Club's rehearsal was changed from Wednesday to Thursday. Or the westside bus would be 20 minutes late. Such announcements were preceded with, "We are sorry to interrupt, but..."

When complaints became troublesome, a committee looked into the matter. For a time, items were more carefully screened, but erosion began all over - another committee. After Henry Gulliver, who installed the sound system, died, it became increasingly difficult to find a repairman who could find the right wire to repair. One day when the entire system sputtered out of control, we discovered that we could live without it. We wondered if there was a saboteur among us.

One of the File 7 memos announced a topic for discussion by the steering committee: "How do we distinguish between a selective skip and an all-day unexcused absence, and why are the penalties different?" Now there is a question! Perhaps it will be answered some day in a doctoral thesis.

Many of the memos are dated only by month and day. I found one that included the year: April 16, 1975. "If you have not already done so, please turn in the form for the NCA study that indicates what percent of the students wear blue jeans to school. These are needed immediately." Now just what did the NCA committee want to do with 700 pairs of blue jeans?

Something must have gone awry on October 27. There was only one announcement on the daily bulletin: "Regular length classes all day."

An undated review of policies listed the following penalty schedule for slow seniors:

3 tardies: notice from vice principal

5 tardies: letter to parents from vice principal

6 tardies: loss of senior privileges

7 to 9 tardies: 30 minute detention

10 tardies: subject to firm administrative action

March 17. A list of 43 students scheduled to welcome the Cedar Rapids Jefferson band. "Don't count them absent from your class." I suspect that more than half of the papers that irritated me were lists of students who would be absent for a "legitimate" activity. Why isn't being in class legitimate?

The May 5 minutes of the steering council were mainly concerned with a top administrative request that teachers log their use of time in certain situations. Marv Scott was quoted: "We're in an impossible bind. If the School Board asks us to file a report because they don't trust us, how can they believe the report?"

December 2: "These students will attend a shoplifting clinic from 8:00 a.m. to 9:50 a.m." Twenty-eight names. Next year a clinic on mafia techniques?

After awhile I stopped reading and merely estimated the dimensions of the file. Nine millimeters thick, 340 grams. Actually, that is a conservative use of paper for topics of such magnitude.

If the eight superintendents and seven principals who contributed to File 7 were to use space equal to mine, they could devastate me in short order. Yes, I confess. There was considerable time when I was not counting blue jeans or releasing students to eradicate shoplifters. I was allowed to teach, encouraged to teach. And, because I was permitted to teach as I thought best, I didn't do it as well as I knew how - I was having too much fun.

I am most grateful for that freedom.

Weekend

It seems to have arrived too soon – but it has been forty-one years.

MOVING OUT

When I hauled my personal things away from Ames High School on that final Friday, I left a number of fragments behind. The cardboard silhouettes of predatory birds, life-size, are still on the wall above the chalkboard: the kestrel on the right, the bald eagle on the left. They look big in a classroom. One first day a boy noticed the birds, raised his arms, aiming an imaginary gun at the eagle. He closed his left eye, and his shoulder jerked back with the recoil. Lowering his arms, he said, "Got 'im!" Later in the day, another boy inspected the silhouettes, shifting his gaze back and forth along the wall. He said, "There's only one Buteo. Is it a redtail?"

I hope my successor likes those birds.

And I hope he or she likes the boxelder stump in the corner of the lab. It's the one with several horseshoes embedded in the wood. A lot of students have puzzled how the shoes got there and how long it took the tree to grow around the steel. In one conversation, I heard a girl explain, "I guess a horse took off his shoes one night and hung them over a limb; next morning he went off barefoot."

There are the hills of prairie west of the building. Squaw Creek, the three colonies of bees, the courtyard pond where we watched the mating antics of the half-wild mallards?

All such bits of biology I dislike leaving. But I am glad they are there.

ANOTHER MONDAY - ANOTHER FALL

Before my "weekend" was old, I accepted a position teaching about bees at Iowa State University. Monday was different from that first Monday 41 years ago; the custodians didn't mistake me for a freshman. The late days of summer, however, haven't changed. I still feel the symptoms of fall fever.

Afterward

Dad taught a few classes in beekeeping at Iowa State University for another ten years, during that time he also assisted with exhibit design at the Living History Farms in Des Moines - another way of teaching. And still, today, at ninety-six he continues writing and revising. Still teaching?.

DMT September, 2008

About the Author

Richard F. Trump first learned nature and the complexity of life in the north-east corner of Missouri and south-east corner of Iowa. As a farm boy of six, near Kahoka, Missouri, as a high-school student in Keokuk, Iowa, and as a college student at Iowa State in Ames, then as a teacher in the Keokuk and Ames high schools, he continued to be a student of nature. But what he learned, he taught. And what he taught, he wrote about.

"Mister Trump" taught biology for 41 years – interrupted only by two years in the U.S. Navy during the Second World War. If his students had known that he was journaling his teaching experiences perhaps they might have been more careful of their antics. But then we wouldn't have as many funny stories.

Dad published two high school biology texts, a delightful exploration of "Bees and their Keepers" and hundreds of articles and columns about everything from the behavior of land crabs in Panama to the wanderings of woodchucks in Iowa. This is the story of his career as a teacher of biology, of human nature and the behavior of high school sophomores, teachers and administrators. He was a keen observer and wrote about what he observed with wry humor and often undisguised delight.

On November first of 2010, he died at the age of 98. (He called it "Checking out.") I'm sure there was still a piece of typing paper in his old IBM Selectric and beside it a spiral-bound book of hand-written notes. He loved to write, and we are richer for being able to read what he wrote. I write this as we celebrate the 100th anniversary of his birth. We miss him.

dmt March 19, 2012

Discovery (a sneak peak)

While teaching biology to high school sophomores was my Dad's official profession, naturalist and writer was his obsession. Dad prowled the fields and woods of north-eastern Missouri as a child and later south-eastern Iowa and central Iowa as a teacher with some time to explore. He didn't explore the Himalayas or the rain forests of the Amazon basin, he explored the wilds that can be found in a back-yard, a park, a farm woodland. And he wrote about his exploration. His journals and writings of his exploration and discoveries in the rural Midwest are chronicled in "Discovery – Field Notes of a Naturalist. Here is a sample:

It's a Snake!

I have found a way to introduce kids to snakes, or snakes to kids. Actually there are no age limits. It works for kids my age. A medium size bull snake will do, or a large garter snake. Place it in a cloth bag, the kind that once held five pounds of sugar. Tie it securely.

"We have a problem," you say. "We must find out what's in this bag without opening it - just by feeling it. Who's first?" It's better not to wait for a volunteer. The longer you wait, the more reluctant they become. Just hand it to someone at close range - girl, a boy, a mom, a dad. It takes only a split second of handling: "It's a snake!" probably accompanied by a shriek,

Reach out for the bag, or more likely - pick it up from the ground. Hold it with both hands, kneading as though you were making a snowball: "Its harmless. Everyone should feel it. Squeeze it and feel its muscles."

Only the braver ones will step forward. But as the timid ones see that the survival rate is high, they too will participate. Eventually you can open the bag. Hold the snake by the neck, and allow it the comfort of coiling around your arm rather than letting it dangle. Then you can show the hikers how to count the rows of scales around the body. The bull snake has 29 to 35, the most for any Iowa native, the blue racer only 17. Within limits such as in the bull snake, the number of rows is consistent for a species and therefore helpful in identification.

The texture of the scales is also helpful. On some species each scale has a small ridge. This gives the skin a rough appearance, while those with smooth scales look slick. Unless they are in a wet environment, all snakes have a dry skin. Touching it is much like touching a leather belt; and it is about the same temperature as the environment.

Some bright kid is going to say, "If that snake isn't poisonous, why are you holding it by the neck?" You could reply, "For the same reason that you are careful when you pick up a thistle. Little teeth can scratch just as little thorns can. Scratches can become infected." Actually most snakes can be trusted not to bite if you allow them the security of coiling around your arm. And some individuals are so trustworthy that you can handle them freely.

The conversation will turn eventually to some of the favorite myths about these reptiles. Will a blue racer chase you? No. Do little snakes crawl down their mother's mouth for protection? No. What kind of snake puts its tail in its mouth and rolls like a hoop? None....

Although I grew up with the usual fears and misconceptions about snakes, I had matured and had found that knowing snakes was helpful in teaching biology.

A snake in the park was sure to follow. At one of my den sites I pushed over a portion of decaying log, to steer a woodchuck into a trap. The snake that I uncovered was truly a miniature, dark back, smooth scales, and a pink belly. Its eyes were small, even for a snake that was no larger that a pencil. Assuming it was a young of a common species, I tried to count the scale rows. Thirteen? I must have miscounted. Fifteen was the fewest of any familiar species. Another count. Thirteen! Back at the hotel, I checked J. E. Guthrie's Snakes of Iowa, Bulletin 239 of the Agricultural Experiment Station.

The worm snake, Guthrie said, had never been reported for Iowa, although it should be here. Well, it was here, in Hotel Manning. That would have been just fine if Carphophis amoena vermis had not escaped. We searched. We sifted our possessions. And we watched furtively as we walked through the parlor, up and down the stairs. We thought it unwise, however, to be seen moving the piano or pulling back carpets. It wouldn't be convincing to say that I'd lost a collar button; I didn't wear collar buttons. My only consolation was that Steamboat Gothic Hotels have lots of leaks. With only thirteen rows of smooth scales, it must have found a way out.

We toyed with the notion that if Bertha Myers had learned that she was housing such a reptile, she would have removed the furniture from the fire escape and posted instructions for evacuation in case of emergency. One long scream followed by three short ones, we thought, would serve as an alarm.

Find Discovery by browsing to AniLibre.com. There you will find links to it in the Amazon online bookstore. It is available both in printed and Kindle format.

Made in the USA
Monee, IL
04 November 2026

67377672R00125